Future of Business and Finance

The Future of Business and Finance book series features professional works aimed at defining, analyzing, and charting the future trends in these fields. The focus is mainly on strategic directions, technological advances, challenges and solutions which may affect the way we do business tomorrow, including the future of sustainability and governance practices. Mainly written by practitioners, consultants and academic thinkers, the books are intended to spark and inform further discussions and developments.

Kevin R. Lowell

Leading Modern Technology Teams in Complex Times

Applying the Principles of the Agile Manifesto

Kevin R. Lowell
Santa Rosa Beach, FL, USA

ISSN 2662-2467   ISSN 2662-2475 (electronic)
Future of Business and Finance
ISBN 978-3-031-36428-0   ISBN 978-3-031-36429-7 (eBook)
https://doi.org/10.1007/978-3-031-36429-7

© The Editor(s) (if applicable) and The Author(s), under exclusive license to Springer Nature Switzerland AG 2023

This work is subject to copyright. All rights are solely and exclusively licensed by the Publisher, whether the whole or part of the material is concerned, specifically the rights of translation, reprinting, reuse of illustrations, recitation, broadcasting, reproduction on microfilms or in any other physical way, and transmission or information storage and retrieval, electronic adaptation, computer software, or by similar or dissimilar methodology now known or hereafter developed.

The use of general descriptive names, registered names, trademarks, service marks, etc. in this publication does not imply, even in the absence of a specific statement, that such names are exempt from the relevant protective laws and regulations and therefore free for general use.

The publisher, the authors, and the editors are safe to assume that the advice and information in this book are believed to be true and accurate at the date of publication. Neither the publisher nor the authors or the editors give a warranty, expressed or implied, with respect to the material contained herein or for any errors or omissions that may have been made. The publisher remains neutral with regard to jurisdictional claims in published maps and institutional affiliations.

This Springer imprint is published by the registered company Springer Nature Switzerland AG
The registered company address is: Gewerbestrasse 11, 6330 Cham, Switzerland

Paper in this product is recyclable.

First and foremost, to my wife, Chris: thank you. To our journey.

To Anna and Jessica: Moon and stars and sun and back.

Preface

Midway in my leadership journey, I found myself wanting to share my view of leadership because I believe that in this life, it's not what you get, it's what you give that matters.

The challenge of leadership is at the same time common and sublime. Most of us lead in one capacity or another, at one time or another, in one circumstance or another, in the journey of our lives. It is our great good fortune to be chosen to experience the humbling, invigorating, daunting, and rewarding challenge of leading others.

This book is a guide to leading modern technology teams in a complex world. I've based it on the experience of other leaders, and on my own experience leading technology teams for nearly 30 years. I asked technology leaders to share their stories, and I include some of those stories here in the form of short vignettes. I also include the findings and conclusions from scholars who research and study leadership. Both sources—scholars and other leaders—provide insights into modern leadership, and both sources continue to inform my own thinking.

I hope you find this book useful.

Santa Rosa Beach, FL, USA   Kevin R. Lowell

Acknowledgments

Thank you to the great teachers and mentors and role models who have inspired me and supported me and enabled me and given me a chance to find myself and to find my way to contribute my verse. The writer Anne Michaels wrote that "the best teacher lodges an intent not in the mind but in the heart." We are beyond fortunate if we have had great teachers. I ask that you in turn be a great teacher for someone else.

To the leaders I spoke with: Thank you for sharing your stories with me. There is power in storytelling. It remains my great good fortune to have heard many stories, including: Chris O'Leary and his story of how to show vulnerability. Deirdre Drake and her story about learning the value of mutually positive outcomes. Clint Wallin and his story demonstrating the power of faith and trust. They shared how they learned from their leaders. Now I've been able to learn from them. And I hope that you find something you can learn from.

Each leader I spoke with had the benefit of a leader who invested in them. I know this because they told me their stories. I ask that you share your story. After all, who are we, if not our stories?

About This Book and How It Works

About This Book

The role of the leader is to create an environment where people can do great work in service of something bigger than themselves. This requires that leadership today is less leader-focused and unidirectional, and more relationship-focused and bidirectional. Leadership today is less about rules than rules of thumb. Modern leadership, and leadership of modern technology teams, is a **social construct** that is **conjunctive**, **processual**, and **generative**. It is creating and causing connections, but it is not alchemy. Leaders in today's technology teams are attentive and deliberate, but not scripted. Leaders are prepared, but not rehearsed. Leadership is a process, not a discrete event. It isn't improv, but it isn't orchestrated, either. For my purposes, the answer to the question "What is leadership?" is this: Leadership is a social construct that is conjunctive, processual, and generative.

This is not a book of theories, but of a framework, of heuristics, integrating and adapting and applying tenets of current leadership practices and scholarship to today's technology teams. This book is not for the scholar. It is for the practitioner who is both well-intentioned and well-informed. Leadership is not about you. Leadership is what you do and how you do it for others. And the leader is you.

Last, the opinions of the leaders in this book are wholly their own. Their opinions should not be taken to represent their respective company's views or opinions.

My opinions are, as you would imagine, my own.

How This Book Works

Each section of this book starts with an overview to summarize the key ideas that will be discussed. This will guide you through the sections that are most relevant to you in your leadership journey.

Part I: The Role of the Leader

Part I defines the role of the leader. It does not dismiss yesterday's leadership models; instead, it offers an approach for leading in today's environment. This part also provides a perspective on complexity.

Part II: Leadership in Today's Technology Organizations

Part II is all about leading. This section describes modern leadership practice as a social construct that is conjunctive, processual, and generative. It breaks down what "conjunctive" means in practice, what "processual" means in practice, and what "generative" means in practice. Part II continues with an explanation of *why* conjunctive and processual and generative leadership practices are necessary in today's complex and changing environment.

Part III: Why Change the Way We Lead?

Part III makes the case for change and is a call to action. It answers the questions, "Why change the way we lead? And change to *what?*" This section frames today's leadership challenge within the context of the changing social and technological conditions. The social changes reflect changes in today's workers and what they care about. It addresses how to lead given today's employees' changing expectations of what they want from their work and what they want from you as their leader. The changing technological conditions include changes in collaboration technologies and information flows—the way work gets done and where it gets done.

Part IV: The Twelve Principles of the Agile Manifesto

Part IV includes practices from employees, leaders, and companies across industries that practice agile software development, or ASD. It answers the questions that you as the leader are asking: "What do I do, and how do I do it?" Think of these as very brief and very targeted documentary biographies not of the broader lives of others but of their specific work experiences. Part IV gets to the nitty-gritty of what other leaders of technology teams are doing today, and of what you as the leader of a modern technology team need to do and identifies ways for you to do it. Part IV includes vignettes of leaders from UScellular, TDS Telecom, Oracle, Amazon, Nokia, Calstate Management Group, Inc., and Centene Corporation. Part IV uses as its organizing framework the twelve Principles behind the Agile Manifesto and presents questions, considerations, and leadership best practices specific to each principle.

Each agile principle has its own chapter and can stand on its own. These can be read in any order. You can read them straight through, or you can read one, skip ahead and read another, then come back later and read another.

The structure of each is the same. Each chapter follows this structure:
"What does the principle mean?"

Here, I describe simply and directly what the principle means. I don't provide an unassailable or definitive definition. Rather, I offer one way to think about each principle. This description in turn sets the stage for understanding, for questioning, for learning, and for doing.

"Let's deconstruct the principle."

Here, I deconstruct the principle. This deconstruction takes each principle apart and looks at the operative words and phrases in each. The words matter. I've found this approach helpful in getting very clear on what the principle really means. A clear understanding lends itself to clear thinking, and clear thinking in turn lends itself to purposeful action.

You Are the Leader: What Do You Do?

Here, I describe the three steps that you as the leader must take. You may prefer a variation to these, or you may have a slightly different approach, or you may take a different tack, or perhaps you have only a notional idea. I encourage you to take these three steps—start with the end in mind, listen and learn from others, and *do*—and to make these steps a mandatory practice in your leadership. Know these three steps. Live by these three steps. Lead by these three steps:

Start with the End in Mind

This means getting clear on where you are heading, getting clear on what you aim to achieve, getting clear on why it matters, and getting clear on what success will look like. Without a plan, without a roadmap, any road will do. To paraphrase Lewis Carroll's Cheshire Cat in *Alice in Wonderland*: If you don't know where you're going, any road will get you there.

Listen and Learn from Others

How do you get to the "end in mind"? How do you get your team on track to achieve its objective? You start by asking great questions, and then you listen. Get very clear on what you are listening for: are you listening for specific answers? Are you listening for tone? Are you listening for aspiration? Are you listening for feedback? Are you listening for reassurance? Are you listening for confirmation? Getting clear on what you're listening for shows respect for the people you're asking, and it makes good use of your time.

Ask great questions, and then listen.

Why spend time up front asking these questions, when you've got work to do and schedules to meet? Two reasons: 1. You don't know everything, and 2. Everyone has a piece of the truth. This includes listening to every member of your team and every one of your customers. They may not have the whole truth, but they have a piece, and it is your job as the leader to seek that out. You do this by listening. Ask great questions and listen to the answers. Your team knows what it does and how it does it. They know what they want, and they know what they need. Your customers know what they want, and they have a perspective on you, your team, and the value of the products that you deliver.

Here's another reason to listen. Consider this from Mike Irizarry, who is the CTO and Head of Technology at UScellular:

Consider:

"Leaders need to listen and when they think they've listened, they need to listen more. And they need to continue to listen, listen to their team, listen to their leader, listen to peers, listen to what is said. And just as importantly, what is not said? Often that is where you find the strength of the team and the opportunities for development. You'll learn what is important to them individually and as a group."

"As I've gone on as a leader, it's less about providing answers, which we often think that's our role as leaders, especially when you're a young leader. But the real trick is the art of the question and figuring out how to ask the right set of questions."

What questions do you ask, and why do you ask them?

I recommend two sets of questions. This first set, below, is what you should ask as you begin working with your team and with your business owner. These are especially helpful if you are new to the team, or if you are undertaking a large new project. More on these below. The second set of questions is specific to each principle, and you'll find these at the beginning of the chapters on each principle.

What questions do you ask first? I recommend these five questions.

1. "What matters most?"
2. "Why does this matter?"
3. "How do you think about X?"
4. "How should I think about X?"
5. "How can I help?"

Let's break down each question.

Question 1: "What matters most?" This question welcomes and encourages aspiration. It invites dreaming. It sets the stage for lofty thinking. And it also stokes a sense of urgency. You get to hear what your team most cares about, maybe even what they're desperate for. Similar for your customer: What must they have? What do they absolutely require above all else? Ask, then listen carefully to their responses.

Question 2: "Why does this matter?" Asking "why?" demonstrates that you are curious and that you are humble. Curious, because you want to know more. Humble, because you are acknowledging that you do not know it all and are willing, even eager, to admit that.

Ask honestly. Do not ask seeking to confirm your biases. Do not ask in the Socratic style, where you are leading the person to the answer you want to hear. Ask honestly and ask humbly. Ask, then listen carefully to their responses.

Question 3: "How do you think about X?" What is the value of asking this question? For starters, you are seeking to understand a different point of view. Asking "how?" after you've asked "why?" encourages wider-ranging responses. "Why?" implies that you are seeking a specific, clinical, direct, and concise answer. Play this out for a moment: if you ask someone *why* they think the way they do, you might prompt a defensive response. When you ask *how*, you invite a description. "How?" implies that you want to understand context and perspective and considerations. You're encouraging a more free-flowing response than "why?" allows for. The question "Why?" requires an explanation. The question "How?" invites a description. Ask, then listen carefully to their responses.

Question 4: "How should I think about X?" With this question, you invite them in. You make it safe for them to tell you how to think and what to do. You encourage dialogue. You initiate a process.

Consider:

"Spend time reflecting on what you bring to the table in light of what you learn in your first few weeks and months with the team, and then be ready to admit that you know you need to make adjustments. I reflect on my own career. You come in thinking you know it. You think that because you're in that role, you must be a great leader. But every situation is unique, and you really need to spend time listening and understanding. Listening and being willing to adjust."

"There's nothing more powerful than sharing your limitations with the team and asking for help. A lot of times we don't want to do that because we think it's a sign of weakness. But I've learned that rather than thinking that you're showing weakness, you're showing confidence and humility."

"Humility isn't weakness. It's really thinking less about yourself and your needs and more about the team's needs and the company's needs. You subordinate your own needs and ambitions to the needs of others in the company. For a new leader especially, that's a hard thing to do. But I would say it's the highest form of respect and honor you can display to someone and the team and the most powerful" (Interview with Mike Irizarry).

Without this invitation, some people will be reluctant to share with you or tell you directly. When you ask the question this way, you're demonstrating humility and curiosity. For you as the leader, these are two of the most powerful behaviors you can demonstrate. Ask, then listen carefully to their responses.

Taken together, Questions 3 and 4 enable you to create a safe space. You are creating psychological safety for people who may be lower in the hierarchy and perhaps reluctant, or who might be intimidated by your positional authority.

Question 5: "How can I help?" Ours is to serve. It's not what you get, it's what you give. You are the leader. You wear many hats. One of your roles as the leader is to find ways to help. This can be by providing tools, clear direction, resources, compassion. The point is you as the leader are here to serve others.

Here I include vignettes from other leaders who have faced challenges like the challenges that you face. These leaders work at different levels in different companies and come from a variety of industries. What they have in common is the challenge of leading modern technology teams.

The specific end in mind, plus the careful and intentional listening, plus learnings from the lessons of others, plus a service mindset, equals a clarity of thought and of purpose. These will inform and guide your actions as the leader.

Do

Do. This is where I tie it together—the answers to the questions you've asked, new questions to ask that are specific to the task at hand, what you've really *heard* in those answers, and how leaders in similar situations think and what they've done. Together, these become a guideline, a roadmap even, but not a blueprint. This is intended to educate, inform, and encourage, but not to dictate. I will recommend *directionally* what you should do and how you should do it. If you are simply going to copy some other leader, to mimic them and imitate them, then clearly you are not ready to lead. But if you really hear what they're saying—both what is said and what is unsaid—and can learn from the lessons of others, then you are ready to do the hard work of leading. Listen and learn, and then get to work. In the words of Michelangelo, "I have been impressed with the urgency of doing. Knowing is not enough: we must apply. Being willing is not enough: we must do."

Contents

Part I The Role of the Leader

1 The Role of the Leader 3
   Today 3
   Yesterday 4

2 A Few Words About Complexity 5
   References 9

3 To the Manifesto! 11

4 Using the Agile Manifesto as a Framework 13
   References 18

Part II Leadership in Today's Technology Organizations

5 What Is Leadership? 21
   Leadership Is a Social Construct 22
   Leadership Is Conjunctive 23
      What Is Conjunctive Leadership? 23
      Why Does Conjunctive Leadership Matter? 24
      What Does the Leader Do? 24
   Leadership Is Processual 25
      What Is Processual Leadership? 25
      Why Does Processual Leadership Matter? 25
      What Does the Leader Do? 26
   Leadership Is Generative 28
      What Is Generative Leadership? 28
      Why Does Generative Leadership Matter? 28
      What Does the Leader Do? 29
   References 30

Part III Why Change the Way We Lead?

6 Employee Expectations Are Changing 35
   Who They Are 35
   What They Do 36

	What They Care About	36
	What They Want in Their Work	37
	What They Want From You	38
	References	39
7	**The Way Work Gets Done Is Changing**	**41**
	Leading in a Hybrid Work Model	41
	Reference	44

Part IV The Twelve Principles of the Agile Manifesto

8	**Agile Principle 1: "Our Highest Priority Is to Satisfy the Customer Through Early and Continuous Delivery of Valuable Software"**	**49**
	What Does This Principle Mean?	50
	You Are the Leader. What Do You Do?	51
	Start with the End in Mind	51
	Listen and Learn from Others	51
	Are You Ready?	56
	Do	57
	Key Takeaways	57
	Reference	57
9	**Agile Principle 2: "Welcome Changing Requirements, Even Late in Development. Agile Processes Harness Change for the Customer's Competitive Advantage"**	**59**
	What Does This Principle Mean?	59
	You Are the Leader. What Do You Do?	60
	Start with the End in Mind	60
	Listen and Learn from Others	61
	Are You Ready?	63
	Do	64
	Key Takeaways	65
	References	65
10	**Agile Principle 3: "Deliver Working Software Frequently, from a Couple of Weeks to a Couple of Months, with a Preference to the Shorter Timescale"**	**67**
	What Does This Principle Mean?	67
	You Are the Leader. What Do You Do?	68
	Start with the End in Mind	68
	Listen and Learn from Others	68
	Are You Ready?	71
	Do	71
	Key Takeaways	73
	References	73

11	**Agile Principle 4: "Businesspeople and Developers Must Work Together Daily Throughout the Project"**	75
	What Does This Principle Mean?	75
	You Are the Leader. What Do You Do?	76
	Start with the End in Mind	76
	Listen and Learn from Others	76
	Are You Ready?	82
	Do	82
	Key Takeaways	86
	References	86
12	**Agile Principle 5: "Build Projects Around Motivated Individuals. Give Them the Environment and Support They Need and Trust Them to Get the Job Done"**	87
	What Does This Principle Mean?	87
	You Are the Leader. What Do You Do?	88
	Start with the End in Mind	88
	Listen and Learn from Others	88
	Are You Ready?	99
	Do	99
	Key Takeaways	119
	References	119
13	**Agile Principle 6: "The Most Efficient and Effective Method of Conveying Information to and Within a Development Team Is Face-to-Face Conversation"**	121
	What Does This Principle Mean?	121
	You Are the Leader. What Do You Do?	122
	Start with the End in Mind	122
	Listen and Learn from Others	123
	Are You Ready?	125
	Do	125
	Key Takeaways	127
	References	127
14	**Agile Principle 7: "Working Software Is the Primary Measure of Progress"**	129
	What Does This Principle Mean?	129
	You Are the Leader. What Do You Do?	130
	Start with the End in Mind	130
	Listen and Learn from Others	130
	Are You Ready?	133
	Do	134
	Key Takeaways	134
	Reference	134

15	Agile Principle 8: "Agile Processes Promote Sustainable Development. The Sponsors, Developers, and Users Should Be Able to Maintain a Constant Pace Indefinitely"	135
	What Does This Principle Mean?	135
	You Are the Leader. What Do You Do?	137
	Start with the End in Mind	137
	Listen and Learn from Others	137
	Are You Ready?	141
	Do	141
	Key Takeaways	144
	References	144
16	Agile Principle 9: "Continuous Attention to Technical Excellence and Good Design Enhances Agility"	145
	What Does This Principle Mean?	145
	You Are the Leader. What Do You Do?	146
	Start with the End in Mind	146
	Listen and Learn from Others	146
	Are You Ready?	150
	Do	150
	Key Takeaways	151
17	Agile Principle 10: "Simplicity—The Art of Maximizing the Amount of Work Not Done—Is Essential"	153
	What Does This Principle Mean?	153
	You Are the Leader. What Do You Do?	154
	Start with the End in Mind	154
	Listen and Learn from Others	154
	Are You Ready?	157
	Do	157
	Key Takeaways	158
	Reference	158
18	Agile Principle 11: "The Best Architectures, Requirements, and Designs Emerge from Self-Organizing Teams"	159
	What Does This Principle Mean?	159
	You Are the Leader. What Do You Do?	160
	Start with the End in Mind	160
	Listen and Learn from Others	160
	Are You Ready?	165
	Do	165
	Key Takeaways	168
	Reference	168

19 Agile Principle 12: "At Regular Intervals, the Team Reflects on How to Become More Effective, Then Tunes and Adjusts Its Behavior Accordingly" 169
What Does This Principle Mean? 169
You Are the Leader. What Do You Do? 170
Start with the End in Mind 170
Listen and Learn from Others 170
Are You Ready? 172
Do 172
Key Takeaways 174
References 174

Part V What's Next

20 Conclusion 177
Bibliography 178

Index 181

About the Author

Kevin R. Lowell's leadership experience spans nearly 30 years. He has lead teams in several disciplines, from retail store design and buildout to website design and operation; from cybersecurity for enterprise and carrier networks, to infrastructure buildouts of the latest wireless technologies; and from the development, delivery, and operation of real-time rating and billing engines and point-of-sale systems, to determining industry standards for the wireless telecommunications.

Today, he is Executive Vice President, Chief People Officer, and Head of Communications at UScellular. His prior roles include Senior Vice President of Information Technology and Vice President of Engineering and Network Operations.

He was named AITP/SIM Chicago 2022 CIO Innovator of the Year. He is a four-time winner of UScellular's Dynamic Leader Award and a five-time finalist for ChicagoCIO's CIO of the Year Award.

Kevin earned his Bachelor of Arts from UCLA, graduating with Departmental Highest Honors in English Literature. He has a Master of Arts Degree in Human and Organizational Systems and holds a PhD in Organizational Development and Change, both from Fielding Graduate University.

His other books include *Make Change! Because Your Life Is Up to You*, a how-to guide for making changes in your personal life; and *In the In-Between*, a love story.

Part I
The Role of the Leader

The Role of the Leader

Abstract

The role of the leader is to create an environment where people can do great work in service of something bigger than themselves. As leaders, this is our *obligation*. The practice of leadership today requires that we start by listening: to our teams, to our customers, and to our colleagues. Yesterday's leadership practices should not be dismissed; instead, we must take the best of the best, apply what we've learned, test and learn more, then move forward. Keep moving forward. Listen, learn, and do.

The role of the leader is to create an environment where people can do great work in service of something bigger than themselves.

Today

Everyone has greatness in them. Our role as leaders—our *obligation* as leaders—is to create an environment where people can do great work in service of something bigger than themselves. We do this by listening. We do this by caring. We do this by trusting. We do this by remaining curious and humble.

You are not the first to ask, "What do I do? How do I lead? What am I supposed to do, and how am I supposed to do it?" You are not the first. You are not alone. And you will not be the last.

This book is about your role. You are the leader. This book is about you and the role you are being paid to perform. This book aims to inform, instruct, challenge, guide, persuade. You are the leader. In your role, you get to choose, you get to decide, you get to direct.

You might be asking, "Why do we need another book on leadership? Aren't there already enough leadership books and leadership stories and leadership examples? Why another one? And why now?"

Here's why: I have written this book not because the times have changed—which they have, without question—but because the way work gets done has changed, and because the expectations of the people doing the work have changed. I don't make the argument that the world of work and the environment in which we live are any more or any less complex than in days gone by. I *do* make the argument that the world of work and the environment in which we live are *differently* complex. Different complexity means different challenges, and different challenges require different solutions. And work done differently requires leadership done differently.

Yesterday

Yesterday's leadership practices should not be dismissed. Nor should they be applied to technology teams without careful consideration. From leadership on the assembly line to leadership in newly electrified offices; from leadership during Depression-era scarcity to leadership during war-time booms; and from leadership in the era of the organization man to leadership in the era of the individual, the practice of leadership continues to evolve in ways assumed to best suit the times. In that regard, today is no different. Leaders always seek to lead in ways that make a positive difference.

Rather than relegating yesterday's practices to the dustbin, I extract what is most relevant, and I adapt those practices to today's complex environment. I share stories from today's leaders on how they lead and what they do, and I apply current thinking from the current research on leadership.

Leadership in the moment is imperfect, imprecise, and sometimes messy, much more so than most leadership books would have us believe. To be clear, this is not a book of theories, but of a framework, heuristics, integrating and adapting, and applying tenets of current leadership practices and scholarship to today's technology teams.

This book is not for the scholar. It is for the practitioner who is both well-intentioned and well-informed. Leadership is not about you. Leadership is what you do and how you do it for others. In this book, the practitioner is the leader. And that leader is you.

A Few Words About Complexity

Abstract

Complexity. The world we live in today is complex. The world yesterday, to the people who lived yesterday, was also complex. Tomorrow will be complex, and so will the day after that. The world has been, continues to be, and always will be complex. Our challenges are as vexing to us as the challenges of the eighteenth, nineteenth, and twentieth centuries were to the people who lived and worked and lived and died in those days gone by.

What distinguishes today from earlier eras is not that we live in times that are *more* complex; rather, we live in times that are *differently* complex. Our challenge as leaders of modern technology teams is not one of facing complexity for the first time. Our challenge is that the way work gets done today is *differently* complex than the way work used to get done. The way work gets done today can be complex; agile software development is *differently* complex. And different challenges require different leadership.

Consider:

"The modern world is brain-splitting in its complexity." Chances are, you have thought this, you might have heard someone say this, maybe you've said it yourself. Maybe not in these specific words, but chances are that you or your colleagues or your team or your leader or your customer has said something similar. We're certain that this applies to us and the world we're living in. We *know* it, and we tend to believe it applies *only* to us. "Clearly, *certainly*," we say to ourselves, "it does not apply to those simple people who came in the years and decades before us, living their simple lives in that simple world of yesterday."

We assume that today's world is *more* complex than ever, and that no one has ever in the history of the world had it as tough as we have it today. "It is the conceit of each new generation to imagine that the problems it faces are more challenging, more rapid and, yes, more complex than those that arose in earlier times (Hughes, 2014)" (Tourish, 2019). We tell ourselves, in part to console ourselves, and—let's face it, to admire ourselves—that today's world is more challenging, faster-paced, more competitive, and more complex than ever. "How simple, how straightforward, how predictable, and how contained the world surely used to be!" we tell ourselves. We like to think that we have it tougher than yesterday's leaders ever could have imagined. We move faster, we're smarter, we bob, we weave, we're more fleet of foot because the world today is more demanding than it's ever been.

The world is complex, and it is uncertain, without a doubt. "In today's environment, complexity is occurring on multiple levels and across many sectors and contexts. Although many forces are driving it, the underlying factors are greater interconnectivity and redistribution of power resulting from information flows that are allowing people to link up and drive change in unprecedented ways" (Uhl-Bien & Arena, Complexity leadership: Enabling people and organizations for adaptability, 2017). The world is complex, uncertain. But is the world *more* complex, is the world *more* uncertain, than it has been over the past 100 years?

If only that were the case. Such thinking is an example of presentism, where we look through today's lens and believe that the past was more predictable than it was. This thinking denies the uncertainty of the past. The people who came before us were no more able to predict their future than we can predict ours.

"The modern world is brain-splitting in its complexity." Think for a moment about this quote. It's from a leading contributor to political thought, a reporter, and winner of two Pulitzer Prizes. More interesting than knowing who wrote this, is knowing when he wrote it. From this leading thinker came the declaration of a world of "brain-splitting" complexity—*our* world, right? That's what Walter Lippmann thought of his world in his times. And the "modern world" he was writing about was the Western world of the early twentieth century. Lippmann wrote this line in 1914. Not 2014, but 1914 (Lippmann, 1914).

Why, in a book on the Agile Manifesto and leadership in modern times, why quote Walter Lippmann? Because the point I strive to make is not about any relevance of that author to this study, but rather the relevance of that *thinking* to this study. Would anyone working or leading or *living* in this modern world deny its brain-splitting complexity? Not likely, at least not in the Western world. No more and no less than the people living when Lippmann wrote these words.

Lippmann's world was complex. Our world is complex. Our world is *differently* complex. "Complexity is about rich interconnectivity" (Uhl-Bien & Arena, Complexity leadership: Enabling people and organizations for adaptability, 2017). As such, this is where we need to qualify our thinking and qualify our conclusions because we are not the first people to experience complexity. We need to qualify our thinking, and, sequentially, our actions. Is our modern world, complex as we acknowledge it be, any *more* complex than the world yesterday? Last month? Last year? A 100 or 300 years ago?

Consider:

It is not that today's times are unlike past times because they are complex. Today is not different from yesterday solely because we see today in all its complexity. "There is little offered to substantiate this declaration of 'unprecedented' change other than assertion. But rhetoric, alas, is not evidence. The challenges society now faces may be different to those of the past. But are they really more complex?" (Tourish, 2019). The world today is fast moving and characterized by "unpredictable local, regional, national, and international contexts, connected ever more closely by new technologies continually re-shaping communication, competition and collaboration" (Collinson, 2014). Yesterday was complex and today is complex. Today is simply *differently* complex. Consider these examples of complexity in the Western world from the past 250 years:

1776: A new nation called the United States of America was born.

1780s: The Constitution of the United States was drafted by a group of 55 white men working all day 6 days a week for 4 months. Its adoption was championed in a series of papers, collectively referred to as The Federalist Papers, written by Alexander Hamilton, John Jay, and James Madison under the pseudonym Pluribus. It was ratified by the citizens of a new nation, fresh off a revolution.

1760s to 1840: the Industrial Revolution.

1860s: A nation divided engaged in a 4-year civil war, family against family, brother against brother. Four million slaves were freed, a president was assassinated, and the course of a nation was changed forever.

1910s: The First World War. The world changed forever. This war to end all wars claimed 20 million lives and laid the groundwork for the Second World War.

1920s: Electrification of industry. "Electrification's effect was universal and revolutionary…In a mere 50 years, the residential United States underwent a transformation from the home production of heat and light by household members who chopped wood, hauled coal, and tended kerosene lamps to a new era of gas and electricity" (Gordon, 2016).

And on.

And on.

And on.

If this isn't change, I don't know what is. If this isn't *upheaval*, I don't know what is.

Not yet convinced that yesterday experienced its own version of complexity? Let's turn once more to Walter Lippmann:

> In the last thirty years or so American business has been passing through a reorganization so radical that we are just beginning to grasp its meaning. At any rate for those of us who are young today the business world of our grandfathers is a piece of history that we can reconstruct only with the greatest difficulty. We know that the huge corporation, the integrated industry, production for a world market, the network of combinations, pools and agreements have played havoc with the older political economy. The scope of human endeavor is enormously larger, and with it has come, as Graham Wallas says, a general change of social scale. Human thought has had to enlarge its scale in order to meet the situation…The size and intricacy which we have to deal with have done more than anything else, I imagine, to wreck the simple generalizations of our ancestors (Lippmann, 1914). Not 2023. 1914.

The more things change, the more they stay the same.
Fast-forward from Walter Lippmann 50 years:

> [Igor] Ansoff, widely regarded as the father of strategic planning, concluded (in 1965!) that the business environment was becoming increasingly 'turbulent', a change he dated from roughly 1950. Mintzberg's (1994) seminal critique of strategic planning notes many similar assertions from the 1960s, 1970s and 1980s to the effect that the current business environment was somehow more turbulent than what preceded it. This notion also has an intrinsic appeal for management gurus, who thrive by offering to ease a panic about the present that they themselves have partially created. For example, Tom Peters (1994) argues that 'crazy times call for crazy organizations' and urged what he called a form of 'perpetual revolution.'… Since the interconnected nature of the challenges that we all face are more evident to us than those that confronted our predecessors forty or fifty years ago it is natural to assume that they are more 'complex'. The strength of this belief doesn't make it true. (Tourish, 2019)

Smith, et al., suggest that "the twenty-first century arguably brought with it unprecedented complexity, diversity and pace to our modern world—globalization, the diffusion of information technology and changing consumption patterns forced organizations to grapple with new or evolving tensions" (Smith et al., 2017). At the same time, could we not argue that the Industrial Revolution brought with it unprecedented complexity? What about the invention of the printing press? Luther's 95 Theses? Spector offers a way for us to think about our modern world without discounting the past. He presents a point of view on presentism and the tranquility fallacy, the terms he uses to label "the tendency to find the current era to be exceptionally, even uniquely turbulent and past eras to seem calm in comparison" (Spector, 2014). "The attribution of turbulence to the present, *presentism,* is typically accompanied by a complementary tendency to attribute tranquility to the past, *the tranquility fallacy*…. When we assume that the present is a time of unprecedented turbulence in comparison to a tranquil past, we constrain our capacity to learn from past experience" (Spector, 2014).

If we discount the past and its challenges, we miss a rich source of learning. "If the world really is now more volatile, unpredictable, complex, and ambiguous (VUCA) than any time in the past, perhaps there isn't time to reflect on issues such as the core purposes pursued by business. Nor is there the need to study other periods of turbulence, and perhaps draw lessons from them" (Tourish, 2019). The very term VUCA, Tourish writes, "is itself an instance of hyperbole. That aside, the suggestion that the Cold War, when we frequently trembled on the brink of nuclear Armageddon, was somehow less complex than what followed it may amuse historians of the period" (Tourish, 2019).

The challenge isn't that there is a new phenomenon called "complexity." Rather, the challenge is that the way work gets done today—specifically and for our purposes, the way that software development gets done—is *differently* complex than the way other types of work get done. Work can be complex; agile software development is *differently* complex.

The notion of complexity, then, is not that it is a novel concept. History has shown us that. Rather, as Rosenhead, et al., state, "complexity offers a potentially valuable metaphor for leadership practice and research. This metaphor holds the possibility of conveying the intricacies and tensions generated in milieux of radical indeterminacy in which, nevertheless, organizations need to take action. Such a metaphorizing of leadership complexity marks a less ambitious—but more practical and grounded role for complexity leadership" (Rosenhead et al., 2019).

As leaders and thinkers, we must guard against the notion that E.P. Thompson expressed 60 years ago, that of "the enormous condescension of posterity" (Thompson, 1966). It is indeed condescending to think that yesterday's leadership challenge involved nothing more than directing a workforce of butchers and bakers and candlestick-makers. Yesterday's challenges were yesterday's challenges; today's, today's.

Think about leaders not as *more* knowledgeable but as *differently* knowledgeable. Consider the world we live in as *differently* complex; think of industry and work and workers as *differently* complex; think of the leadership challenge—*your* challenge—as *differently* complex. We're challenged differently, so we need to think differently.

What does any of this have to do with the Agile Manifesto? Read on.

The Agile Manifesto is a twenty-first-century creation. It was a response to the emerging and challenging environment of software development. It is to today's world of agile software development what Taylorism was to early-twentieth-century process engineering. It prescribes the conditions for working productively, not in assembling automobiles but in building software. Taylor defined the principles of scientific management. The Agile Manifesto defines the Principles of software development.

The work of software development did not exist 100 years ago. Leadership challenges, on the other hand, *did* exist 100 years ago. Leadership, whether inspired by Isaac Newton or Frederick Taylor or William H. Whyte or butterflies, is fluid. There's the leadership orthodoxy, then there's leadership in action: the changing, adapting, adjusting. Leadership isn't sterile and academic. If leadership were a place, it would not be an operating room. It would be a craftsman's workshop or an artist's studio.

I'm less concerned in this book with what's testable than with what's practical and applicable and valuable. There is no single approach, no right answer. Rather there are angles of approach, there are options. And that's okay. Think of it as "cognitive refinement" of all the practices that have come before. The scientific method is its own ontology, and that method may or may not be suitable here. I'm most interested in applying old and new ideas to improve the quality of leadership and organizational outcomes.

References

Collinson, D. (2014). Dichotomies, dialectics and dilemmas: New directions for critical leadership studies. *Leadership*, 10, 36–55.
Gordon, R. (2016). *The rise and fall of American growth*. Princeton University Press.
Hughes, M. (2014). Leading changes: Why transformation efforts fail. *Leadership*, 12, 449–469.

Lippmann, W. (1914). *Drift and mastery: An attempt to diagnose the current unrest*. Mitchell Kennerley.

Rosenhead, J., Franco, L. A., Grint, K., & Friedland, B. (2019). Complexity theory and leadership practice: A review, a critique, and some recommendations. *The Leadership Quarterly*, 1–25.

Smith, W., Erez, M. J., Lewis, M., & Tracey, P. (2017). Adding complexity to theories of paradox, tensions, and dualities of innovation and change: Introduction to the special issue on paradox, tensions, and dualities of innovation and change. *Organization Studies*, 303–317.

Spector, B. (2014). Using history ahistorically: Presentism and the tranquility fallacy. *Management and Organizational History*, 305–313.

Thompson, E. (1966). *The making of the English working class*. Penguin Books.

Tourish, D. (2019). Is complexity leadership theory complex enough? A critical appraisal, some modifications and suggestions for further research. *Organization Studies*, 219–238.

Uhl-Bien, M., & Arena, M. (2017). Complexity leadership: Enabling people and organizations for adaptability. *Organizational Dynamics*, 9–20.

To the Manifesto! 3

Abstract

Work done differently requires leadership done differently. The authors of the Agile Manifesto recognized that the work of software development was a different kind of work. The purpose of the Manifesto was to guide this work, enable this work, and assist the people doing this work—the software developers—to do the work to the best of their abilities.

Work done differently requires leadership done differently. The authors of the Agile Manifesto recognized that the work of software development was a different kind of work. The purpose of the Manifesto was to guide this work, enable this work, and assist the people doing this work—the software developers—to do the work to the best of their abilities.

Rather than trying to reinvent corporate America, let's do the best with what we have. We can talk and theorize and speculate and pontificate all day long. But I'm with Michelangelo, who said, "I'm impressed with doing." Let's get something *done*.

The times, they are complex. The way work gets done is different. Agile software development is different. On to the Manifesto!

Using the Agile Manifesto as a Framework

Abstract

The Manifesto for Agile Software Development, or as it's commonly referred to, the Agile Manifesto, lays out the values and principles whose aim is to enable software development teams to work more efficiently and effectively in service of their customers.

Think of the values as the guiding principles for agile software development and delivery. Then think of the principles as structuring and informing the practice of agile software development and delivery.

The values emphasize the importance of individuals and their interactions. The values emphasize that making progress matters more than making something perfect. Collaboration matters. Responding to changes—in priorities, in needs, in requirements—matters.

The principles are not rules, and they are not strictures. Instead, they are tenets that structure and inform the practice. The principles provide a framework, and they are ours and they are yours to apply in service of your team and in service of your customers.

For convenience, I've numbered each Principle, and I'll refer to the Principle either directly or by its number. They're numbered in the order they originally appeared.

Manifesto for Agile Software Development

We are uncovering better ways of developing software by doing it and helping others do it. Through this work we have come to value:

Individuals and interactions over processes and tools
Working software over comprehensive documentation
Customer collaboration over contract negotiation
Responding to change over following a plan

That is, while there is value in the items on
the right, we value the items on the left more.

Kent Beck	James Grenning	Robert C. Martin
Mike Beedle	Jim Highsmith	Steve Mellor
Arie van Bennekum	Andrew Hunt	Ken Schwaber
Alistair Cockburn	Ron Jeffries	Jeff Sutherland
Ward Cunningham	Jon Kern	Dave Thomas
Martin Fowler	Brian Marick	

The Twelve Principles of the Agile Manifesto

We follow these principles:

1. Our highest priority is to satisfy the customer through early and continuous delivery of valuable software.

2. Welcome changing requirements, even late in development. Agile processes harness change for the customer's competitive advantage.

3. Deliver working software frequently, from a couple of weeks to a couple of months, with a preference to the shorter timescale.

4. Businesspeople and developers must work together daily throughout the project.

5. Build projects around motivated individuals. Give them the environment and support they need and trust them to get the job done.

6. The most efficient and effective method of conveying information to and within a development team is face-to-face conversation.

7. Working software is the primary measure of progress.

8. Agile processes promote sustainable development. The sponsors, developers, and users should be able to maintain a constant pace indefinitely.

9. Continuous attention to technical excellence and good design enhances agility.

10. Simplicity—the art of maximizing the amount of work not done—is essential.

11. The best architectures, requirements, and designs emerge from self-organizing teams.

12. At regular intervals, the team reflects on how to become more effective, then tunes and adjusts its behavior accordingly.

What exactly is "agile software development"? Agile is a delivery methodology. Agile is a way to get work done. It "refers to a group of approaches to software development using iterative (repeated processes) and incremental (successively added functionality) development" (KnowledgeHut Solutions Private Limited, 2011-23). The Agile Alliance defines it this way: Agile is "the ability to create and respond to change. It is a way of dealing with, and ultimately succeeding in, an uncertain and turbulent environment." There are different applications of Agile (think SAFe for implementing Agile practices at scale) and different methods of Agile (Scrum, Extreme Programming (XP), Kanban, Feature-Driven Development). Dinakar Hituvalli, Group Vice President for product development at Oracle, says this of Agile: its value is in the speed of development and the ability to react quickly to changes.

While there are different Agile methodologies, the importance of good leadership remains unchanged. Effective leadership is necessary across all of these. This book focuses on effective leadership in the context of the 12 Principles of the Agile Manifesto.

Let's think about the four values of the Manifesto:

1. Individuals and interactions over processes and tools
2. Working software over comprehensive documentation
3. Customer collaboration over contract negotiation
4. Responding to change over following a plan

1. Individuals and interactions over processes and tools

What does this mean?

This means "people first." This means trusting the person over the process, vision over Visio, faith over flowcharts, creativity and spontaneity over command and control. This means trust and leaps of faith when otherwise we demand evidence. It's belief despite proof. This means encouraging and empowering. This means believing in what could be.

As the leader, when you put your trust in the individual and in interactions over processes, you're trusting in the generative nature of dialogue. You're trusting that something better, something new, something different, will emerge from the dialogue and the interactions between people and among groups. What more empowering statement to make to a person than to say, "I trust you"?

This is an example of where leadership is processual. The interaction between you and your employee is a process: it's dialogue, it's back-and-forth, give-and-take. It's co-creation. New ideas, new ways of thinking, new possibilities, and new solutions all can emerge from open and honest dialogue. And almost certainly, even *especially*, when that dialogue takes place between two people who start out disagreeing. Elie Wiesel, Holocaust survivor and Nobel laureate, shared this example: "There is a wonderful Hasidic teaching about this that says when two people disagree, and each one pulls away to his own side, his own opinion, a space is created between them. In this space, worlds can be created, provided the two antagonists do not fill the space with too many words. It is only because two people disagree that there can be such a space; were they to hold identical positions, there would be no room for innovation. In other words, conflict can be a good thing—if it is done well" (Burger, 2018).

2. Working software over comprehensive documentation

What does this mean?
This means don't talk about it, do it. This means make something happen. This means that you value things that work over things that describe what might be in a perfect world. It also means let's not overthink things. Let's get something done. It's a bias for action. Get something down and have something to work with.

Muhannad Obeidat is a leader at Oracle. He described that in his 20+ years of experience in many different roles, the teams "used to spend a ton of time building functional requirements, then building technical requirements, then getting into designs. Today, we don't do that. We still build a good amount of documentation, but we don't spend a lot of time on documentation before getting started."

Muhannad's practice is consistent with a practice described by complexity theory. This theory posits that comprehensive strategic plans should be replaced with simple documents describing a general direction for the future of an organization augmented with a few basic principles on how best to get there (Zimmerman et al., 2008). Detailed planning is not considered to be a good use of time for leaders because the external environment changes rapidly and is unpredictable.

"We build what we call writeups, which are very short documents that describe user goals, user stories, and functional requirements, and then immediately jump into demo-able scenarios. We start experimenting, doing proof of concepts, then writing technical designs as we go along. We don't have that big cycle of weeks or months going by just sending documentation back and forth before we build anything."

What is the value of documentation? It enables reuse without rework. It also helps minimize technical debt. Documentation that cannot be overlooked is the documentation on how the software works, and the requirements that the software was built to deliver on.

The objective isn't documentation. The objective is working software that delivers value. Delivering valuable software is the name of the game.

3. Customer collaboration over contract negotiation

What does this mean?
This means that what your team does is not about them. Their work is done in service of others.

Productive collaboration requires positive relationships. Positive relationships become a source of power, enabling organizations to evolve and adapt because the people in them care more about their work, their co-workers, and their shared purpose. A culture of caring emerges where people are willing to go to great lengths not to let others down. The resultant behaviors lead to greater creativity, innovation, and productivity (Zimmerman et al., 2008). When people are connected to others through a shared purpose for the success of an organization, they often become capable of doing more than initially envisioned (Kotter, 2012). They are willing to contribute more to meet the needs of the organization which leads to greater feelings of personal fulfillment.

In terms of the Manifesto, it means that we work with others to understand others. We understand others so we can direct our efforts to meeting their expectations. This means we prioritize others over ourselves.

4. Responding to change over following a plan

What does this mean?

This means that you as the leader need to achieve balance between delivering on the original plan and innovating to accommodate change. The effective leader engages her team in responding to changes. Your role as the leader is twofold: (1) create the environment in which your team understands its accountabilities and (2) get out of the way and let the team do what it does best.

The degree of change in today's organizations is best embraced by a leadership style that accommodates and even encourages change. Systems theorists believed that managers should strive for greater alignment between an organization and its environment to achieve optimal performance (Simon, 1996). But more recently, complexity theorists suggest that in view of the rapid change in today's systems, a more promising approach is for leaders to acknowledge the need for continual change and cultivate the optimal conditions for change to occur.

Leaders do this "by creating space for ideas advanced by entrepreneurial leaders to engage in tension with the operational system and generate innovations that scale into the system to meet the adaptive needs of the organization and its environment...This becomes a critical form of leadership for adaptive organizations" (Uhl-Bien & Arena, 2018).

Next, let's turn to the principles. The word "principle" is a noun. Two definitions from Oxford Languages: (1) a fundamental truth or proposition that serves as the foundation for a system of belief or behavior or for a chain of reasoning; (2) a general scientific theorem or law that has numerous special applications across a wide field.

Each Principle in the Agile Manifesto has a specific meaning. Words have definitions. We cannot change the definition of the word simply because we want it to mean something different. If we start changing what words *mean*, then we don't have a grounding or a foundation for *any* meaning. Each means what it means. Nothing will mean anything because everything will mean anything.

The 12 Principles of the Agile Manifesto guide and structure and inform the practice of agile software development. The 12 Principles for ASD "help establish the tenets of the agile mindset. They are not a set of rules for practicing agile, but a handful of principles to help instill agile thinking" (ProductPlan, 2023).

We'll break down each Principle in Part IV. I will explain what each one means in practice, and I will describe how they can be applied in today's competitive environment. I do this not to discuss whether we need to change each Principle. Rather, I break down each principle to map it to the current day and our current environment. We must understand each, and we understand them by breaking them down. We need to understand the meaning of each, then determine its suitability in our current complex times. We determine how we might apply them to our work. To determine this, we need to understand each Principle. Each may turn out to be directly applicable; some may turn out to be extensible. We're not here to change the principles but rather to challenge them and find what's best in them and determine how we might apply them to our current complex times.

The Principles are not right or wrong; they simply *are*, and they are ours to work with. We need to understand them, and then apply them or extend them or emphasize or deemphasize them as we lead our teams. We get to decide what we want to do and how we want to do it. You are the leader. You have the latitude and the obligation as the leader of a technical team in today's complex environment to determine how best to apply each.

References

Burger, A. (2018). *Witness: Lessons from Elie Wiesel's classroom*. Houghton Mifflin Harcourt.

KnowledgeHut Solutions Private Limited. (2011-23). *knowledgehut.com*. Retrieved from knowledgehut.com: www.knowledgehut.com

Kotter, J. (2012). *Leading change*. Harvard University Press.

ProductPlan. (2023). *www.productplan.com*. Retrieved from ProductPlan web site: www.productplan.com

Simon, H. (1996). *The sciences of the artificial*. MIT Press.

Uhl-Bien, M., & Arena, M. (2018). Leadership for organizational adaptability: A theoretical synthesis and integrative framework. *The Leadership Quarterly*. https://doi.org/10.1016/j.leaqua.2017.12.009

Zimmerman, B., Lindberg, C., & Plsek, P. (2008). *Edgeware: Lessons from complexity science for health care leaders* (2nd ed.). VHA, Incorporated.

Part II

Leadership in Today's Technology Organizations

Abstract

Leadership today needs to be less leader-focused and unidirectional and more relationship-focused and bidirectional. It's less about rules than rules of thumb. Modern leadership, and leadership of modern technology teams, is a *social construct* that is *conjunctive*, *processual*, and *generative*. It is creating and causing connections, but it is not alchemy. Leadership in today's technology teams is attentive and deliberate but not scripted. It is prepared but not rehearsed. It is a process and not a discrete event. It isn't improv, but it isn't orchestrated, either.

For my purposes, the answer to the question "What is leadership?" will be answered directly and succinctly:

Leadership is a social construct that is conjunctive, processual, and generative.

What Is Leadership? 5

Abstract

Leadership today is a social construct. It evolves in response to demands and changes in the social, political, economic, and business environment. Leadership is practiced in the relationship between a person designated as the leader and the person, team, division, or organization that person has been designated to lead.

In practice, leadership is conjunctive, processual, and generative.

Conjunctive Leadership. Conjunctive leadership connects people with other people and connects people with resources, ideas, and objectives. Vision is the catalyst of conjunctive leadership. The leader who can see and imagine and then connect ideas and learnings from separate domains is demonstrating conjunctive thinking. Conjunctive leadership leverages this interconnectedness. Conjunctive leadership matters because no single person or single team accomplishes the objectives of the organization or meets the needs of the customer alone. No team is an island, and no one succeeds alone.

Processual Leadership. Processual leadership is a series of actions. It is not a discrete event. Processual leadership is a fluid process, not a fixed state. It is through the process of dialogue that includes diverse thinking that the best ideas and the best solutions emerge. In this process, leaders engage others, they listen, they seek to understand, and then they act.

Generative Leadership. Generative dialogue can be thought of as the practice of identifying corollaries. Corollaries include combinations, transference, and borderlands between and across bodies of knowledge, and they lie at the heart of innovation.

Generative leadership matters not only because today's challenges are global and unprecedented in our lifetimes but also because our current state changes rapidly. The world we live in and the world we work in changes rapidly and changes at scale. It demands more than one thinker. It demands more than one voice. It demands inclusive dialogue.

Leadership Is a Social Construct

Leadership is negotiated between and among members of an organization. In its practical application, leadership is a process in the fluid relationship between a person designated as a leader and a person or persons designated as individual contributors, a team, or another leader.

Effective leadership has changed over the past 20 years. We've shifted from a leadership model that was leader-focused and unidirectional, to a practice that is conjunctive and processual, more generative, more adaptive. This is what leadership *is:* conjunctive, processual, and generative.

Leadership is not a title, and it is not defined by a box on an organization chart. Leadership can be applied down, as organizations have traditionally practiced it, but it can also be applied across and even up.

Consider these definitions of leadership:

"The formal definition of leadership as I thought about it is the practice of pulling together talent resources in the most efficient and effective way to enable a company to execute its vision, mission and purpose through the company's more tactical, go-to-market strategy. I'd say that leadership is as important to a company as the nervous system is to a human being. If it's conditioned well and in good health, then companies can execute their vision and mission in a very effective way, leading to shareholder value and everything that comes along with that" (Interview with Mike Irizarry, 12/19/22).

Leadership is uniting "those groups, those teams, those individuals to pursue something and achieve it with passion and energy and positiveness. Leadership transitioned from being much more directive and authoritative, in my view, to much more collaborative and much more united." (Interview with Cece Stewart, 12/19/22).

"I think about leadership as the ability to enable people under your leadership to accomplish whatever it is they're trying to accomplish. How? Removing obstacles and enabling people to thrive" (Interview with Deirdre Drake, 12/21/22).

"Leadership is being able to influence people to follow you and to focus on and buy into the goals that you want to accomplish" (Interview with Sam Crowley, 12/20/22).

And this from Clint Wallin, the Senior Director of Platforms and Delivery Services at UScellular: "Leadership is the ability to influence and help people achieve a goal or an objective."

That is what leadership is. But what do leaders *do*? When leaders are leading, what exactly is it that they are doing?

A leader thinks broadly and acts directly. A leader considers the dynamic of the company, of the industry, of the customer, and of the environment, including social and cultural dynamics. A leader communicates. A leader focuses. A leader is effective in the face of ambiguity. A leader develops talent. A leader gets results. Leaders "typically exercise considerable control over scarce resources; decision making; structures, rules and regulations; formal communications; strategies and visions; corporate culture; performance management; rewards and sanctions; and hiring and firing" (Collinson, 2014). Followers apply the resources, execute on decisions, and do the work consistent with the direction set by the leader and consistent with the norms of the culture.

Dennis Tourish has written extensively about leadership. "Good leadership develops strategies that address compelling needs, takes advantage of opportunities for change, and builds a coalition of willing followers prepared to work collectively in pursuit of common goals" (Tourish, 2020).

The binding element between leaders and followers is trust. "Trust is the currency of leadership. Without it, all you're left with is authoritarian leadership, and that's not sustainable. Your team is not going to stick around if they don't trust you" (interview with Mike Irizarry 12/19/22). Say what you mean and do what you say. This should be the credo of every leader. When you say what you mean, you demonstrate that you are transparent. When you do what you say, you demonstrate that you are credible and trustworthy. Trust and credibility are bedrocks of the social construct that is leadership. A leader says what she means, and she does what she says.

Leadership Is Conjunctive

What Is Conjunctive Leadership?

Conjunctive leadership connects people with other people and connects people with resources, ideas, and objectives. The leader who can see and imagine and connect ideas and learnings from separate domains is demonstrating conjunctive thinking. Consider the example of Franklin D. Roosevelt. Doris Kearns Goodwin, in *Leadership in Turbulent Times*, describes Roosevelt's early life "revealing a unique transverse intelligence that cut naturally across categories" that for him became "a characteristic mode of problem solving" (p. 57) (Kearns Goodwin, 2018). Disconnected fields, connected to create new ideas.

Think of Einstein's notion of combinatorial play, that "process of considering two or more unrelated ideas, topics, images, disciplines, etc. and putting them together in a way that is new" (I Teach University, 2016). It is the "act of opening up one mental channel by dabbling in another." He considered it "the essential feature in productive thought." Think of Arthur Koestler's Theory of Bisociation, which is a deliberate linking or "bisociation (not mere association) of two (or more) apparently incompatible frames of thought" (Kirkmann, 2006). Conjunctive thinking is characterized by "seek[ing] to make connections between diverse elements of human experience through making those distinctions that will enable the joining up of concepts normally used in a compartmentalized manner" (Tsoukas, 2017).

How do you do this? Look across departments, look across domains, look across the industry, look at different industries. Then tease out the interconnectedness. I encourage you to read the profiles of Nobel Laureates in *Cultures of Creativity*. Many of the Laureates describe their process of creativity. First, they thought big: "What's possible, what do I believe in, where do I want to go?" And then they thought small: "What already exists, what is in front of me, what is available to me?" What are the possibilities—think of these as dots—and what might appear or occur if you connect them?

Vision is the catalyst of conjunctive leadership. To know where to look and what to connect, you as the leader must have a vision of what you believe is possible, where you want to take your team, and what resources are available to you. This is forward-looking. This means that you are anticipating and seeing and causing deliberate connections for the creation of something new. This requires that you see possibilities and allow for new and unanticipated outcomes.

Why Does Conjunctive Leadership Matter?

Conjunctive leadership matters because no single person or single team accomplishes the objectives of the organization or meets the needs of the customer alone. No team is an island. No one succeeds alone. Period.

What Does the Leader Do?

The best leaders lead with humility. "The only wisdom we can hope to acquire is the wisdom of humility. Humility is endless" (Eliot, 1943). It is not about the leader; it is about the success of the team. It is not about a heroic, front-page, headline-grabbing personality; it is about the success of the team. The leader's role includes building, enabling, empowering, trusting, and appreciating the team. Today's leadership "is often much less hands-on and much more behind the scenes than traditional leadership. It is also more distributed, involving sharing credit and working collaboratively, rather than hierarchically. Therefore, it can, and often does, go unrecognized in organizational systems that focus on strong, hierarchical forms of leadership" (Uhl-Bien & Arena, 2018). Today's leaders "must be convicted enough in what they are doing to take great risks in opening up adaptive space for others, and humble enough to step back so others can step forward" (Uhl-Bien & Arena, 2017).

The best leaders connect to communicate. The best leaders connect with their customers to better understand their customers. The best leaders connect with their teams to share, learn, guide, direct, challenge, and develop. One behavior of the best leaders is that "they ask more questions than they give direction. They do more listening than talking" (Interview with Doug Lowell, 7/21/22).

Communicate *how*?

Clearly and honestly, without flourish and without hyperbole.

Communicate *what*?

The best leaders know that *what* you communicate depends on your objective. Chris O'Leary was formerly the interim CEO, Tupperware, and former EVP and COO of General Mills. He was responsible for $7 billion of revenue, leading 30,000 global employees. He emphasizes the importance of communication and the importance of being clear on what you are communicating. In Chris's experience, "The currency is communication. But what are you communicating? It depends. Sometimes, leaders are trying to inspire and motivate. Sometimes they're coaching. Sometimes they're simplifying. Sometimes they're prioritizing and making decisions. But it all comes through the currency of communication."

Communicate *why*?

The best leaders listen to learn and seek to understand. Learn and understand your company's mission. Learn and understand your customer. Learn and understand your team. Then, connect them. With this understanding, you can connect your team to the company's mission and to your customer. "I would say that the secret sauce for leadership is connecting the individual and their team's notion of meaningful work to the company's mission and strategy" (Interview with Mike Irizarry). "Strategic leadership forges a bridge between the past, the present, and the future, by reaffirming core values and identity to ensure continuity and integrity as the organization struggles with the known and unknown realities and possibilities (Boal, 2004 quoted in Boal & Schultz, 2007, 412)" (Rosenhead et al., 2019).

Leadership Is Processual

What Is Processual Leadership?

Processual leadership is a series of actions. It is not a discrete event. Processual leadership is a fluid process, not a fixed state. "It is universally accepted that organizations can be seen as both an entity and a process (Langley et al., 2013)" (Tsoukas, 2017). As such, leadership is never "done." It is, according to Schweiger et al., "a processual perspective [that] views leadership as an ongoing social interaction involving all organizational actors" (Schweiger et al., 2020). It is "a social interaction process that is co-constructed by all involved actors... [The] practical implications from a processual perspective include a focus on interactions and reciprocal influences [where leaders] are keen to take a change in perspectives, recognize productive potential of dissent, ask questions to others, [and] are inherently interested in others' points of view" (Schweiger et al., 2020).

The leader creates the forums and the opportunities for dialogue and this engagement. These forums range from deliberately constructed public events to informal, unplanned one-to-one interactions. These forums and opportunities can include team meetings, mid-week socials, open forums (more on open forums later in Principle 5), town halls, mentoring and reverse mentoring (more on these in Principle 5), and innovation fairs. "Because leadership is always collectively enacted in situation, it becomes a consequence of actors' relations, an effect processually generated by a group of people, a product of their local interactions" (Denis et al., 2012).

The construct is less important than the interaction. The interaction is what matters.

Why Does Processual Leadership Matter?

We need processual leadership now more than ever because our challenges are differently complex. No one individual can solve for the simultaneous challenges and the difficulties posed by a worldwide pandemic, a war in Europe, climate changes, social upheaval, and a looming recession. Harry Harczak, Jr., adds to this list today's unprecedented reliance on global supply chains and on technology. He describes this current environment as "chaotic." The magnitude of the challenges and the fact that we haven't had to confront simultaneous challenges of this magnitude ever before *demands* more than one mind thinking through things. We need more minds engaged. We need more voices in the dialogue. We need different perspectives.

Bringing this global leadership challenge closer to home and closer to our workplace, the value of many minds and many voices involved in establishing a path forward is no less important. The business challenges of today, coupled with the global challenges of today, require the inclusion of many minds and many voices engaged in dialogue. And dialogue is, by definition, an event that takes place over time. Dialogue is an exchange, and an exchange is a process. Conjunctive leadership brings the minds and the voices and the people and the perspectives together; processual leadership engages them in dialogue. And next, generative leadership makes things happen.

What Does the Leader Do?

The best leaders think broadly and act directly.

The best leaders consider the dynamic of the company, of the industry, of the customer, of the broader environment and of the immediate environment, including social and cultural dynamics, and then they act. A leader understands the dynamic of the business, the importance of the objective, and the role of the team well enough to determine the next best actions.

Sam Crowley underscores the idea that leadership is situational. "You're going to be a different kind of leader if you're on the Titanic and the boat's going down. You're trying to save as many people as possible. You haven't got time to convince anybody that the way you're leading them is the right way. You just have to make it happen. You are going to make it happen now, and people need to understand that. Of course, every now and then you'll get some resistance, but most people tend to respond well to a leader with the right methodology. You have to be decisive. You have to respond to the situation, and if people have questions, you need to listen enough to understand. But you can't spend a lot of time with analysis." Leaders act.

The best leaders bring people together, and they engage with them. These leaders work to establish an organizational routine of connecting with customers, their peers, and their teams. They engage resources and connect them—see *Leadership is conjunctive* above—and they enable new and unexpected or unanticipated or even unimagined outcomes.

The best leaders provide the organizing framework—daily stand-up meetings, utilization of the kaizen approach, or a Kanban board—a description of the value to be created, and the accountability that encourages employees to produce. Dionysiou and Tsoukas describe that such "organizational routines are an important element of organizational behavior" (Dionysiou & Tsoukas, 2013). Tourish writes that "leadership cannot be meaningfully depicted as a force that stands apart from complex systems, neutrally exerting influence and control to achieve putatively positive outcomes" (Tourish, 2019).

In the process of bringing people together, one of the roles of the leader is to listen. "I urge that we collectively pay much more attention to what Alvesson and Sveningsson (2003) described as the small and even mundane acts whereby leaders perform leadership and seek legitimacy, such as merely listening and chatting to others" (Tourish, 2019).

The best leaders also interpret situations. The best leaders interpret the competitive environment, they interpret resource constraints, and they interpret potential futures. They are not stunted by ambiguity. Rather, they understand the situation well enough to think clearly and communicate clearly.

One way that the best leaders make interpretations of complex or ambiguous situations understandable for their teams is metaphors. The best leaders develop metaphorical thinking and recognize the value of metaphors. Metaphorical descriptions can provide employees with new perspectives and nuanced points of view that in turn aid in their thinking and their understanding.

Take, for example, this especially effective metaphor. Laurent (LT) Therivel, CEO of UScellular, applied a hotel-room metaphor when he was describing to a mostly nontechnical audience the abstract notion of capacity of a wireless network for handling customer phone calls and customer data sessions. He asked the audience to think about an example that was easy to understand: the number of rooms in a hotel. Hotels are big buildings full of rooms. When the hotel opens on its first day, it has dozens or hundreds of rooms available for guests to use. Then the rooms start to fill up. The total capacity that the hotel is capable of is fixed because there is a fixed number of rooms. When people start to book rooms and stay at the hotel, the available capacity decreases. The more rooms that are booked and fewer the number of rooms available for the next customer. Similar for wireless network capacity. Networks are built with capacity, and that capacity gets used by people when they make phone calls or use apps to play games or shop or navigate. Wireless network capacity is a challenging notion to conceptualize. Metaphoric thinking makes it easier to understand. This is what leaders do: interpret situations and make them easily understandable for the team.

The best leaders act. They may lead from the front or support from behind. Work in the weeds or at 50,000 feet. Where you lead from depends on what you see. And what you see depends on where you look. From Gruver, Young, and Fulghum:

> First and foremost, we don't see that the manager's job is to tell the organization what and how to do its job in a command-and-control sense documented in some agile books. It is to clearly define the strategic directions and organizational priorities and then, working with the team, to get everyone driving in that direction. Next, it is to monitor progress and work with the teams to understand what is and what isn't working. The managers then help the organization improve where things are struggling. Or it means getting the needed help from another team where dependencies exist. It could be working with the product owners to get clearer definitions for a user story or negotiating for a smaller feature load so they can catch up with some development debt. It could be working with the team to highlight and drive process changes that would make the organization more effective. It could be working with upper management to increase funding or capacity in that area, so they are set up for long-term success. In any case, the role of the management team is to define the direction and then work with the team to help the team succeed. (Gruver et al., 2012)

Processual leadership is not only about creating harmony in the choir. It is also about promoting dissent. "A processual perspective on leadership suggests placing more emphasis on the promotion of dissent and difference than the achievement of harmonious dialogue and consensus (Collinson, 2018; Tourish, 2019)" (Schweiger et al., 2020).

One voice only is called a solo, and no one succeeds alone. And few businesses are one-dimensional. "Leaders deal with contingencies and possibilities rather than linear sequences. Indeterminacy, uncertainty, and unpredictability are ever present and can never be eliminated" (Tourish, 2019). Business problems are multivariate and therefore require multiple inputs and multiple perspectives and multiple iterations to solve them. One of Amazon's leadership principles is appropriate here: disagree, then commit. Or as one of my best bosses told me early in my career, when he sensed that I had a point of view different than his and was reluctant to say so: "If we always agree on everything, I don't need one of us." And it was clear to me which one of us he wouldn't need.

Leadership Is Generative

What Is Generative Leadership?

As the leader, when you put your trust in the individual and in interactions over processes, you're trusting in the generative nature of dialogue. You're trusting that something—something better, something new, something different—will emerge from the dialogue and the interactions between people and among groups. You're enabling these "somethings" to emerge by connecting people in the process of dialogue.

Generative dialogue can be thought of as the practice of identifying corollaries. Corollaries include combinations, transference, and borderlands between and across bodies of knowledge, and it lies at the heart of innovation. Consider this perspective describing the work of Arne Tiselius, winner of the Nobel Prize in Chemistry in 1948: "To create something new and unknown from several known phenomena – the act of making a new combination – lies at the heart of creativity" (Science History Publications/USA & The Nobel Museum, 2007).

Why Does Generative Leadership Matter?

Generative leadership matters not only because today's challenges are global and unprecedented in our lifetimes, but also because our current state can and does change rapidly. There isn't time enough for a heroic leader to remove herself from daily life and daily work and retreat into solitude for days or weeks on end and then emerge with the answer. Even if a leader did remove and retreat to ponder and consider, and then emerged with an answer, it would be an answer to yesterday's question. The world we live in and the world we work in changes rapidly and changes at scale. It demands more than one thinker. It demands more than one voice. It demands inclusive dialogue. Rapidly changing situations—globally, socially, competitively— "require people to act independently far more often, rather than waiting for direction from above. It follows that using the leadership potential of multiple constituencies within organizations is likely to be crucial for success. This means that power and the capacity to make decisions need to be more widely distributed" (Tourish, 2020). How do you as the leader develop this leadership potential and this leadership capacity in your team and throughout your organization? Develop talent, develop talent, develop talent. (But how? Read on. See ideas and examples for developing talent in Principle 5, "Build projects around motivated individuals." Give them the environment and support they need and trust them to get the job done.)

"If we do live in a complex world, it makes more sense to see leaders and followers as interacting organizational actors whose identities as leaders and followers are simultaneously constructed and deconstructed by the force of their ongoing respective struggles to realize their agentic potential (Tourish, 2013)" (Tourish, 2019). "Constructed" and "deconstructed" is the conjunctive aspect of leadership. "Ongoing" is the processual aspect of leadership. And realizing their "agentic potential" is a generative aspect of leadership.

What Does the Leader Do?

The best leaders think broadly and act directly.

The best leaders articulate a clear and compelling vision.

The best leaders model the ethics and the values and the behaviors that the company stands for. "As the leadership goes in this regard, so goes the health of the company" (Interview with Mike Irizarry, 12/19/22).

The best leaders deliver.

It may go without saying, but I will say it anyway: the way leaders get things done is through other people. The best leaders "are being able to separate a process that's not leading anywhere from process that is leading somewhere" (Interview with Doug Lowell, 7/21/22). Busy-ness isn't the measure; valuable results are the measure.

Leadership is conjunctive, processual, and generative. The connection, and then the interaction and the dialogue between you and your employee is a process: it's dialogue, it's back-and-forth, give-and-take. This is a critical moment in the dialogue. You as the leader cannot let it stop here. The dialogue must not stop with each side digging in and anchoring to a position. Rather, this is the point where the dialogue becomes generative. This is co-creation. New ideas, new ways of thinking, new possibilities, and new solutions all can emerge from open and honest dialogue. And almost certainly, even *especially*, when that dialogue takes place between people who start out disagreeing. Recall Elie Wiesel's example using the Hasidic teaching about disagreement creating space for innovation. It is an example of generative dialogue. It's a coming together. It is deliberate work to find a possibility, to find something new. Think of this when businesspeople and product owners talk with your software development team. Think of this during the feedback to a demo. Is the dialogue combative, contentious, argumentative? Or does it allow for something new, something creative? Does it allow for something to emerge?

References

Alvesson, M., & Sveningsson, S. (2003). Managers doing leadership: The extra-ordinarization of the mundane. *Human Relations*, 56, 1435–1459.

Collinson, D. (2014). Dichotomies, dialectics and dilemmas: New directions for critical leadership studies. *Leadership*, 10, 36–55.

Collinson, M. (2018). What's new about Leadership-as-Practice? *Leadership*, 14(3), 363–370. https://doi.org/10.1177/1742715017726879

Denis, J.-L., Langley, A., & Sergi, V. (2012). Leadership in the plural. *The Academy of Management Annals*, 6(1), 211–283. https://doi.org/10.1080/19416520.2012.667612

Dionysiou, D. D., & Tsoukas, H. (2013). Understanding the (re)creation of routines from within: A symbolic interactionist perspective. *Academy of Management Review*, 38(2), 181–205.

Eliot, T. (1943). *Four quartets*. Faber and Faber.

Gruver, G., Young, M., & Fulghum, P. (2012). *A practical approach to large-scale agile development: How HP transformed LaserJet FutureSmart firmware*. Addison-Wesley Professional.

I Teach University. (2016, March 21). *www.iteachu.com*. Retrieved from www.iteachu.com: https://iteachu.uaf.edu

Kearns Goodwin, D. (2018). *Leadership in turbulent times*. Simon & Schuster.

Kirkmann, A. (2006). Contemporary linguistic theories of humour. *Folklore*, 33, 27–57. https://doi.org/10.7592/FEJF2006.33.kriku

Langley, A., Smallman, C., Tsoukas, H., & Van de Ven, A. (2013). Process studies of change in organization and management: Unveiling temporality, activity and flow. *Academy of Management Journal*, 56, 1–13.

Rosenhead, J., Franco, L. A., Grint, K., & Friedland, B. (2019). Complexity theory and leadership practice: A review, a critique, and some recommendations. *The Leadership Quarterly*, 30(5), 1–25.

Schweiger, S., Muller, B., & Guttel, W. H. (2020). Barriers to leadership: Why is it so difficult to abandon the hero? *Leadership*, 16(4), 411–433. https://doi.org/10.1177/1742715020935742

Science History Publications/USA & The Nobel Museum. (2007). *Cultures of creativity: Birth of a 21st century museum*. Watson Publishing International LLP.

Tourish, D. (2013). The dark side of transformational leadership: A critical perspective. *Development and Learning in Organizations*, 28(1).

Tourish, D. (2019). Is complexity leadership theory enough? A critical appraisal, some modifications and suggestions for further research. *Organization Studies*, 40(2), 219–238.

Tourish, D. (2020). The triumph of nonsense in management studies. *Academy of Management Learning & Education*, 19, 1.

Tsoukas, H. (2017). Don't simplify, complexify: From disjunctive to conjunctive theorizing in organization and management studies. *Journal of Management Studies*, 54, 132–153.

Uhl-Bien, M., & Arena, M. (2017). Complexity leadership: Enabling people and organizations for adaptability. *Organizational Dynamics*, 46, 9–20.

Uhl-Bien, M., & Arena, M. (2018). Leadership for organizational adaptability: A theoretical synthesis and integrative framework. *The Leadership Quarterly*, 29, 89–104. https://doi.org/10.1016/j.leaqua.2017.12.009

Part III

Why Change the Way We Lead?

Abstract

This section makes the case for change and is the call to action. It frames today's leadership challenge broadly within the context of changes in the way work gets done and in the changing environmental and social conditions. These changing conditions include changes in employees' expectations of the leadership, their work, and their company, and changes in companies' relationships with the world outside their four walls.

Employees' wants and needs evolve, industries evolve, social environments evolve, responses and solutions evolve. Why wouldn't or shouldn't leadership evolve? There are as many theories, models, approaches, and styles of leading as there are ways of working. Consider that Taylorism saw its greatest influence in the 1910s, during which time steel manufacturing and Henry Ford's assembly lines for automobile manufacturing were dominant modes of production. In this case, there appears to be consistency between the leadership model—the mechanistic and precise Taylorism—and the production model—assembly lines. Yet it was at this same time—the 1910s—that Walter Lippmann, writer and political commentator, declared that in current society, "the scope of human endeavor is enormously larger" and that "the modern world is brain-splitting in its complexity" (Lippmann, 1914). These declarations reappear in much the same way when today's commentators write of our VUCA (volatile, uncertain, complex, and ambiguous) environment.

This book is filling a gap in the "theory to practice" space in a manner suitable to today's technology organization. Leader-follower dynamics are complex—they include power dynamics, conflict, ambiguity, and paradox—and technology organizations are complex. This doesn't mean, however, that all prior modes and models of leadership are inapplicable. Max Weber had theories. Sigmund Freud, Erik Erikson. More recently, Jim Collins, Kouzes and Posner, Daniel Goleman. We've had servant leadership, transformational leadership, situational leadership, contextual leadership, charismatic leadership, and heroic leadership. We've had Lewin's three styles of leadership: autocratic, democratic, and laissez-faire. We've had delegative leadership, participative leadership, bureaucratic leadership, and visionary leadership. We've had hands-off leadership, hands-on leadership, leadership from the front, leadership from behind, and leadership by walking around. Today's challenges do not require that we reinvent leadership. Rather, it is time to extend and evolve yesterday's leadership practices to meet today's leadership challenges.

Think about the cultural tension that you as the leader are up against:

- Big tech
- Layoffs
- Inflation
- The Great Resignation
- Work from home

No worker is 100% satisfied with their leader or with leadership. With this book, I'm not inventing a problem to solve. And we all—you as the leader and the people that you lead—work and live with this multidimensional tension. The problem exists. Which means that *opportunity* exists.

The way work gets done is changing. There's a value shift to cloud solutions and to platforms. There's a continuation and an expansion of the value of networked connections. Not only this, but the wants and needs of the people who are doing this work are changing, too. Last but not least, the relationship between the company and the employee is changing.

This shift is evident both in companies' external brand and employment proposition and in companies' internal orientation and internal communications to their employees. The thinking is that in this age of Great Resignations and Great Regrets and Great Reconsiderations and Great Renegotiations, the company must be focused on the employee to attract and retain employees. And not just *any* employees, but *key talent*, talent that is in demand and talent that can provide differentiated products and differentiated services. This line of thinking continues: a focus on the employee satisfies the employee and frees the employees' mindshare to focus on customers. So rather than making customer satisfaction the stated sole of the company, companies are focusing on their employees, too, who in turn are focusing on the customer. "We take care of you," the company says to the employee, "and you take care of the customer."

As your company culture is only as strong as your commitment to it, same goes for your team: your team is only as strong as your commitment to it.

Leaders today say that their business is much more complicated and less controllable than it used to be. Leaders continue to be surprised at how many decisions turn out to be in error and how many planned interventions fail to achieve their objectives. Complexity theory has increasingly shown that acting based on an understanding of simple linear relationships in a business environment is not enough to ensure success. Further experience and analysis using nonlinear models might enhance our understanding of complex environments and our selection of the optimal approaches for managing organizations. Until then, leaders need to understand what needs to change in their leadership approach, and how to apply these new practices.

One of the challenges will be—and always is—leaders anchoring to what might go wrong and choosing not to change because the change might fail. Leaders too often are afraid of what might not work. They're afraid of the risks, the downside, the uncertainty. They're afraid of failure. We've all witnessed this: leaders who are afraid to take a chance. Leaders who are afraid to change. Leaders are paid to weigh risk, and weighing risks requires that we understand the context for the choices and the decisions that we're making. But it doesn't require that we get stuck. Just the opposite. It means that we must take calculated risks. It means that we must be effective in the face of ambiguity.

You are the leader. Here's one way to deal with this: ask yourself, "What's the worst that can happen?" What's the worst that can happen if you're wrong, or you're late, or your forecast is off, or your prediction misses the mark? Unpleasant things, sure; bad things, maybe. But *disaster*? Probably not. Winston Churchill faced risks of a magnitude few of us will ever face. And yet he was decisive. Was he always right? Of course not. Was he always popular? Not even close. But he was the leader and he made decisions and he kept it all in context. He famously said, "Success isn't final. Failure isn't fatal." Many mistakes are recoverable. Many decisions are reversible.

The world is differently complex as we've established. But let's get more specific: within this differently complex world, employee expectations have changed, and the way work gets done has changed.

Reference

Lippmann, W. (1914). *Drift and mastery: An attempt to diagnose the current unrest*. Mitchell Kennerley.

Employee Expectations Are Changing

Abstract

Today's employee is described in many ways: by age, by generation, by motivation, by hopes and dreams. Regardless of the various descriptions, one attribute is common across today's employees: each is an individual. Each is a unique individual with unique characteristics, both visible and unseen. Each is unique in who they are, what they do, what they care about, what they want in their work, and what they want from you, their leader.

Today's employee wants—and in many cases, demands—a sense of purpose. The desire for a sense of purpose is a fundamental human need. Fulfilling work is more than a title or a salary. Today's employee wants to invest their time, energy, and attention in meaningful work performed in a hybrid environment.

Who are the employees doing this work? Who are these "new" employees? Who are you leading?

Most days do not consist of life-changing moments. This does not mean, though, that you as the leader cannot make the days meaningful. How do you do this? You do this by knowing what is important to you, to your company, and to your team. And how do you learn this? You start by asking questions. And then you listen.

Let's start by describing today's employees. Then we'll describe their expectations.

Who They Are

> I am large, I contain multitudes. – Walt Whitman.

Who are they? They are seven living generations:

1. The Greatest Generation (born 1901–1927)
2. The Silent Generation (born 1928–1945)
3. Baby Boomers (born 1946–1964)
4. Generation X (born 1965–1980)
5. Millennials (born 1981–1995)
6. Generation Z (born 1996–2010)
7. Generation Alpha (born 2011–2025)

Chances are that you are working with four of these generations. What are the common experiences within each generation, and what are the common experiences across these generations? Witness: the fall of the Soviet Union, the moon landing, the assassination of a president, the Cold War, the Vietnam War, the Civil Rights Act, the crumbling of the Berlin Wall, the end of Apartheid, the invention of the personal computer, the invention of the internet, a global pandemic.

These, to name just a few. All of these are visible, tangible, and external. The list of all that goes on *within* us—love, hope, fear, confidence, uncertainty, and more—goes on and on, too. It includes issues of work/life harmony, personal growth, hybrid work models, and diversity in social, ethnic, sexual, and intellectual orientation, and where diversity is a relationship and not a thing or a condition. My list won't be the same as yours, and no list will be exhaustive. "After all, we are not at the end of social evolution ourselves" (Thompson, 1966).

What They Do

What do today's employees do, and how do they do it? They work. They play. They integrate the demands of the day—demands of employers, customers, family, and the self—into 24 hours, then they get up and do it again. They struggle, they aspire, they wrestle. They push forward as best they can. Some feel they move ahead, and some feel they fall behind. Some are well and some are sick. Most want to do good. Most want to connect in a meaningful way with someone else and something more. Most want to feel they can, and most want to feel they matter. This is who you are leading, and this is what they do.

What They Care About

They care about wellness.
    They care about community.
    They care about acceptance.
    They care about continued personal support.
    They care about professional growth.
    They care about clear work/life boundaries.

They care about volunteering.
They care about their development.
They care about society.
They care about the environment.
They care about the social good that the company is doing.
They care about their identity.
They care about their heritage.
They care about equality.
They care about equity.
They care about health.
They care about work/life harmony. (I first heard the recasting of work/life balance as "work/life harmony" from Tim Walter, SVP, Global CIO, CISO, Randa Apparel & Accessories, in an interview we did together on August 3, 2022).
They care about flexibility.
They care about sustainability.
They care about the digital divide.
They care about representation as a strategic imperative.
They care about fairness.

What They Want in Their Work

Today's worker wants a sense of purpose. In fact, most *demand* a sense of purpose in the work that they do. The desire for a sense of purpose extends well beyond titles. The desire for a sense of purpose is a fundamental human need. Who among us does *not* desire a sense of purpose and a reason for being and a reason for doing? Is that desire for purpose and for meaning any different in the soul of the front-line worker or the individual contributor, than it is in the leader? Than it is in you? Today's workers want to invest their time, energy, and attention in meaningful work performed in a hybrid environment. They want a sense of purpose. In that regard, they are no different than you or I. A life lived without a sense of purpose is a rudderless life.

UScellular CEO Laurent Therivel believes that people today think differently about the meaning and the value of work.

Consider:

"First, the notion of work being noble in and of itself is somewhat obsolete. Generally speaking, anybody who wants to work and make money can work and make money. Working and being the traditional 'breadwinner' is available more broadly than ever. Second, capitalism isn't delivering everywhere. Economies are growing in many cases, but at what cost? What about the social cost? What about the environmental cost? What about the cost to the climate? People aren't seeing that capitalism solves for these problems. So, people want more. They want to do meaningful work. The definition of 'meaningful' has changed. It's in the eye of the beholder. It's personal. It has become personal because it can be personal. People have choices. People look around, they're not 100% satisfied with what they see, and they feel empowered because of choice to expect and demand and *do* something else, *do* differently" (Interview 12/19/22).

Consider:

"What do today's employees want? Back when I entered the workforce, back then, you kept your head down. You did your job. You got your check. 'Corporate social responsibility.' What is that? 'Be a good corporate citizen.' What is that?"

"Employers might have done some things because they were a large employer in the area or out of the goodness of their general mission, but it wasn't obligatory. Where now, as an example of generational change, this generation demands more from the corporation. They demand social responsibility. They demand environmental responsibility. They demand more from the companies they choose to work for."

"And companies need to listen because there are so many more potentially viable options for people. Today, they can pursue an entrepreneurial path or pursue a small company path or choose not to engage with the large corporate machine unless the large corporate machine does something to accommodate their needs."

"Each generation that enters the workforce has different expectations based on what they saw happen with their parents, and so they've tried to modify the construct of work to meet their needs and not have the same experience that they saw their parents go through. They demand balance" (Interview with Deirdre Drake, 12/21/22).

What They Want From You

They want empathy.
   They want understanding.
   They want clarity.
   They want compassion.
   They want acceptance.
   They want fairness.
   They want opportunity.
   They want transparency.

They want accountability: for themselves, their co-workers, their leaders, and their company.

They want a leader who understands the value of diversity and is willing to do something about it. "Agility of thought based on the fostering of diversity is a prerequisite for the organization's longer-term success in a wayward environment" (Rosenhead, Franco, Grint, & Friedland, 2019). Diversity is more than race, origin, social background, ethnic background, gender, or sexual orientation, and it is why diverse representation matters. Cece Stewart, the former banking executive, sees diversity this way: "It's diversity of backgrounds and thinking and experiences. Diversity is incredibly important. The best leaders recognize that." Deirdre Drake: Today's workers "demand to come to work and be able to not have to dim any light that they may bring to the table. We've come to appreciate as leaders that those differences are beneficial to us, where in the past homogeneity had been rewarded. I've seen the leadership discourse around the benefit of diversity evolve."

They also want courage and honesty and conviction and humility. They want a leader who is smart. In this world we live in, your teams don't want a showman or a huckster. They don't need a hero. They need a leader they can trust. To paraphrase Stephen Denning in *The Age of Agile*, complexity responds to competence, not authority (Denning, 2018). Today's employees want to be recognized for who they are and for all they are.

To extend Walt Whitman: Workers are large, they contain multitudes.

References

Denning, S. (2018). *The age of agile: How smart companies are transforming the way work gets done*. American Management Association.

Rosenhead, J., Franco, L. A., Grint, K., & Friedland, B. (2019). Complexity theory and leadership practice: A review, a critique, and some recommendations. *The Leadership Quarterly*, 30, 1–25.

Thompson, E. (1966). *The making of the English working class*. Penguin Books.

The Way Work Gets Done Is Changing 7

Abstract

Work gets done differently today: hybrid arrangements, remote arrangements, global teams collaborating across time zones and oceans. Environments are different, customer expectations continue to evolve, and employee expectations continue to evolve.

These differences and these changes require a change in the way we lead. These differences underscore the importance of effective communication. Effective leaders recognize the differences in work environments, and they leverage one or more mediums to communicate effectively.

Leading in a Hybrid Work Model

What does leadership look like in a hybrid work environment? You are the leader. What do you do when many of your interactions with your team are done virtually and not in person? "Leadership in this environment revolves around establishing trust between leadership and their teams" (Interview with Doug Lowell 7/22/22).

Consider:

"Technology has enabled leadership to transcend the traditional boundaries of locality and to increase the surface area of leadership and the potential of leadership to influence. With technologies like Teams and Zoom, you can really influence more broadly and more deeply within your company but also with the company's ecosystem of partners. And if you don't recognize that, you could really do harm. But if you do recognize it, you could really extend the influence of your leadership, and not for selfish purposes, but for really helping your people, the team and the company achieve its objective. So, I don't think that technology has changed the fundamentals of leadership, but it's definitely enabled it and given some more power to it" (Interview with Mike Irizarry, 12/19/22).

Consider:

"What does leadership look like in a hybrid work environment? Well, it has changed drastically from my perspective. I've had to become more deliberate with my communications. I've had to really think about the importance and the need to communicate more frequently. I need to use different channels to communicate. When I have my all-hands meetings, I really think through what I need to communicate. How am I getting people up to speed on different projects that they might hear about if they were sitting next to somebody who was working on that project, but now they're not hearing about it in this hybrid environment? I've had to learn, and I'm still learning.

"There's so much more organic communication that occurs when you've got for example 40 people sitting in an office. We do stay connected in our agile scrum meetings every day and the daily standups, but those daily stand ups are quick and they're very focused on what's immediate: 'What am I doing today? What did I do yesterday? Do I have any challenges with my current work assignment?' It's not about, 'Hey, I heard this might be coming. We need to be preparing for this,' or 'Hey, we've got a new technology that we're thinking about.' In a hybrid environment, there's a lot that's missing that just happens organically in the office. So, we as leaders need to be more deliberate about what might be missing and that we're communicating" (Interview with Jeff Mander 10/9/22).

Jeff's point is a very important one. You as the leader need to be aware of the communication that *isn't* happening. You as the leader need to be aware of the communication that perhaps you take for granted when you and your team are in an office environment, where the teams can walk down the hall, have that quick conversation, and be connected. You as the leader need to be deliberate in your communication so that you can fill in the gaps where important information might be missing.

Alex King helped establish "Amazon Remote," the first Employee Resource Group for virtual employees at Amazon. This fully volunteer organization is committed to creating a positive, supportive, and productive experience for employees as they navigate the virtual work environment. The group focuses on several aspects of the remote work model:

- Improve the experience of remote-working employees by compiling/providing access to information, reducing ambiguity, and empowering community members to drive change within their organizations
- Gather a community in which members can connect with one another in such a large company
- Acknowledge the spectrum of remote work (e.g., style of work and impacts on different identities) and provide guidance applicable to employees in their situation
- Support employees through their entire employment lifecycle, from onboarding to promotion and career development
- Educate managers and individual contributors on the benefits of remote work and provide tools for general inclusion of remote work

What does leadership look like in this environment?

- Modeling and prioritizing collaboration opportunities for the team, both work- and social-related.
- Empowering employees to manage their own time and hours, within reason. Obviously, teams need times available where many people can participate.
- Setting realistic expectations for delivery (e.g., identifying realistic velocity, understanding that 8 hours of 'code writing' per day simply isn't feasible).

"One of the most important parts of leading in this hybrid environment is figuring out how to build personal connections, especially when we have people who are new to the team. Three years into the hybrid environment, and we're still figuring it out. We need to figure out how we make better and stronger personal connections" (Interview with Jeff Mander, 12/21/22).

Building connections requires that people spend time together, virtually or in-person. Being engaged as a leader means being present and being available. "If you're in person, they just walk to your office and you're available, but when your team is working remotely and they're connecting virtually, you have to figure out a way of being available when your team needs you. And that needs to be outside of regularly scheduled meetings. You need to accommodate some ad hoc availability in addition to programmed availability" (Interview with Doug Lowell, 7/22/22).

Your ad hoc and impromptu availability is important for your team. You will need to establish boundaries so that you don't implicitly create an expectation that you are available at all hours every day. At the same time, you as a leader must respect your team's boundaries. Vacation, personal time, family time: your availability and your team's availability need to be discussed and agreed upon. If you don't, you set yourself up for an experience like this one that Doug shared about a woman who had worked for 2 years with no vacation time and then took a Disney vacation with her family. The first day of vacation, walking into the park, her boss sent a text. He had a "quick question," and asked that she get on a "quick call." It went on like that for the rest of the week until she was so frustrated and in fear of her job that she couldn't enjoy a minute of the vacation. She got back and resigned.

Some leaders will knowingly take advantage of their employees. Some managers act out of a lack of concern for the employee's well-being. Setting boundaries and respecting boundaries is not only a reasonable thing to do. It's the respectful thing to do. It's the right thing to do. You must respect boundaries.

You are the leader. How do you do this?

One way to do this is to establish protocols of expectations for how you and your team are going to interact given that the team is remote and working virtually. "Discuss and agree to what kind of response time you should expect from each other to an e-mail or to a phone call. If you or someone on your team needs something urgently, what's the best way to request that? Text them on their mobile? Send an instant message? Phone call? Not only will protocols help you establish and communicate boundaries, but they will also help the team recognize what the level of urgency is for any interaction" (Interview with Doug Lowell, 7/22/22).

Given our experience of remote work and hybrid work arrangement, it appears likely that the future of work will include these arrangements. For many, the traditional 5-days-in-the-office routine has come and gone. Professor Nicholas Bloom shared results from the Survey of Business Uncertainty that he runs with the Atlanta Fed and the University of Chicago. "Before COVID, 5 percent of working days were spent at home. During the pandemic, this increased eightfold to 40 percent a day. And post-pandemic, the number will likely drop to 20 percent. But that 20 percent still represents a fourfold increase of the pre-COVID level, highlighting that working from home is here to stay" (Bloom, 2020). "The US economy is now a working-from-home economy" (Bloom, Han, & Liang, How hybrid working from home works out, 2023).

You as the leader need to recognize this, accept this, and work with your team to establish protocols that will help all of you thrive in this environment.

Reference

Bloom, N. (2020). *How working from home works out*. Stanford Institute for Economic Policy Research (SIEPR).

Part IV

The Twelve Principles of the Agile Manifesto

Abstract

This section (re)introduces the Agile Manifesto and describes how to lead specific to each Principle.

"The technology team must have a client orientation. And the leader must be client oriented. In this way, the technology team can become a competitive differentiator by introducing new thinking and new solutions. Technology teams have the opportunity not only to deliver what the business asks of them and needs from them, but also this opportunity to transform the business" (Interview with Laurent "LT" Therivel). Don't simply answer. Don't simply solve. Answer and *create*. Solve and *create*. This is the expectation that business leaders have of technology leaders and technologists and technology teams: delivery and strategic enablement and strategic transformation. This is the generative aspect to leading a technology team.

Oxford Languages defines a manifesto as "a public declaration of policy and aims." Wikipedia defines it as "a published declaration of the intentions, motives, or views of the issuer, be it an individual, group, political party or government." And Britannica.com defines "manifesto" as "a document publicly declaring the position or program of its issuer. A manifesto advances a set of ideas opinions, or views, but it can also lay out a plan of action." Manifestos, it continues, "are generally written in the name of a group sharing a common perspective, ideology, or purpose rather than in the name of a single individual."

Airfocus defines the Agile Manifesto this way: "The Agile Manifesto is a document that sets out the key values and principles behind the Agile philosophy and serves to help development teams work more efficiently and sustainably." Key values. Key principles. Help. Teams. Efficient and sustainable.

Manifesto for Agile Software Development

We are uncovering better ways of developing
software by doing it and helping others do it.
Through this work we have come to value:

Individuals and interactions over processes and tools
Working software over comprehensive documentation
Customer collaboration over contract negotiation
Responding to change over following a plan

That is, while there is value in the items on
the right, we value the items on the left more.

Kent Beck	James Grenning	Robert C. Martin
Mike Beedle	Jim Highsmith	Steve Mellor
Arie van Bennekum	Andrew Hunt	Ken Schwaber
Alistair Cockburn	Ron Jeffries	Jeff Sutherland
Ward Cunningham	Jon Kern	Dave Thomas
Martin Fowler	Brian Marick	

The Twelve Principles of the Agile Manifesto

We follow these principles:

1. Our highest priority is to satisfy the customer through early and continuous delivery of valuable software.

2. Welcome changing requirements, even late in development. Agile processes harness change for the customer's competitive advantage.

3. Deliver working software frequently, from a couple of weeks to a couple of months, with a preference to the shorter timescale.

4. Businesspeople and developers must work together daily throughout the project.

5. Build projects around motivated individuals. Give them the environment and support they need and trust them to get the job done.

6. The most efficient and effective method of conveying information to and within a development team is face-to-face conversation.

7. Working software is the primary measure of progress.

8. Agile processes promote sustainable development. The sponsors, developers, and users should be able to maintain a constant pace indefinitely.

9. Continuous attention to technical excellence and good design enhances agility.

10. Simplicity—the art of maximizing the amount of work not done—is essential.

11. The best architectures, requirements, and designs emerge from self-organizing teams.

12. At regular intervals, the team reflects on how to become more effective, then tunes and adjusts its behavior accordingly.

Let's dive into leadership and the twelve Principles.

Agile Principle 1: "Our Highest Priority Is to Satisfy the Customer Through Early and Continuous Delivery of Valuable Software"

Abstract

Delivering valuable software frequently is the name of the game. Developers and customers have different definitions of what constitutes value. You learn what "value" means to your customer by bringing her together with your team. You accomplish this through dialogue. The dialogue generates clarity of requirements and a common understanding of the process. Developing software that satisfies the quality, speed, and value expectations of customers is what matters most. This is what customers want and need, and this is what they are paying you for, so this is what you as the leader must deliver. Leaders of agile software development teams know that this is far easier said than done. It involves dialogue, iterative development, risks, and failures.

Vignettes from leaders at Nokia, UScellular, and Centene Corporation illustrate these leadership challenges and how to address them.

It begins with dialogue. You as the leader bring your agile software development team together with the business owner to establish and execute a sustainable, repeatable process for prioritizing requirements, defining what will satisfy the customer and how the customer will use the software solution, and defining what the customer deems valuable. The objective of joining the teams and executing the process is to generate solutions. Leaders overcome challenges by implementing a repeatable and sustainable process for generating solutions. Teams must agree on what "valuable" means and what "continuous" means. Answers aren't revealed. Rather, solutions are generated, and they're generated through dialogue.

What Does This Principle Mean?

What does this Principle mean? Here's a way to think about this: "We care most about the customer. We care most about *satisfying* the customer. We know what satisfies the customer because we listen to the customer."

Let's deconstruct this Principle.

Nothing matters more than satisfying the customer through early and continuous delivery of valuable software. This simple sentence is loaded with key ideas and key commitments. Let's take this principle apart. Let's break it down.

"Our highest priority." Our highest priority is what matters most to our customer. It is our objective. It is our true north. It is our target.

"Satisfy the customer." This principle gives us our target— "satisfy the customer"—and tells us how to do it. This principle makes it clear that customer satisfaction is to be found not in the lowest price or the prettiest package, but in the early and continuous delivery of valuable software. Your customer will not be satisfied with cheapest or slickest. They will like inexpensive, and they might be wowed by "slick." But they have needs, and they expect you to meet those *needs*. Just ask them.

"Early and continuous delivery." What do we mean by "early and continuous delivery"? "Early" in this case means "sooner rather than later," and never late. "Continuous" is on-going, consistent, regular, and dependable. Important, too, to recognize that when you and your team are delivering software continuously, you are working on it continuously. You are adding to it, improving it, modifying it. You're listening to your customer and you're adjusting the software as you go *in ways that are valuable to your customer*. Change for the sake of change is like treading water: you're busy doing something, but you're not getting anywhere.

"Valuable software." You as the leader need to understand from your customer what "valuable" means to them. For certain, they will *not* define "valuable" as "buggy" or "inconsistent" or "unreliable" or "too darn complicated to use." Your customer will not be satisfied with a low price if the product doesn't meet their needs. Your customer will not be satisfied with pretty packaging if the product doesn't meet their needs. Software delivered late is less valuable to your customer than software delivered on time. It's even more valuable if you can deliver it early.

Change the question that you ask your customer from, "What would you like?" to "What matters most?" Then break the work, the project, the program, into small discrete chunks that can be described clearly.

Continuous delivery means continuous development. It means continuous monitoring and continuous measuring. It also means continuous conversations. From Stephen Denning's *The Age of Agile*: "Teams regularly monitor how the feature are being used. The results flow into the aspirational backlogs, which are called scenarios. Every month, the program manager reports out on metrics of the accounts, measuring different aspects of the service. The group is learning to become a 'data-informed' business. They don't call it 'data driven' because that would run the risk of missing the big picture" (Denning, 2018).

You Are the Leader. What Do You Do?

Start with the End in Mind

What will this look like when you've achieved it? When this Principle is in place, your team will be getting iterations of working software into the hands of your customers as quickly and consistently as you can.

Listen and Learn from Others

How do you get to the "end in mind"? You start by asking great questions, and then you listen (see the section "Listen" in "How this book works"). For this principle, I encourage you to ask these questions of your team, your customers, your peers, and your boss:

1. "What matters most to your customer?"
2. "How will your customer use this product or feature?"
3. "What matters most to you?"
4. "Why does this matter?"
5. "How do you define and determine 'customer satisfaction'?"
6. "How do you define 'early and continuous delivery'?"
7. "How do you define 'valuable software'?"
8. "How should I think about what it takes to deliver valuable software? How should I think about what it takes to satisfy the customer?"
9. "How can I help?"

Let's break down each question.
"What matters most to your customer?"
"How will your customer use this product or feature?"
"What matters most to you?"

I asked these questions of Mike Brendzal, Healthcare Technology Leader and Agile Practitioner. Mike has worked on the business side, so comes at this question from the perspective of a business owner who understands employee needs, business needs, and customer needs. Mike is in the uncommon position as a businessperson also having a deep understanding and appreciation of technology capabilities and technology constraints. From Mike's perspective, the focus of the principle, and much more importantly, the focus of the team's *effort*, is on *value*. If you as the leader are not delivering *value* early and continuously, why bother delivering at all? Why would you as the leader guide and direct your team to deliver something, *anything* that is not valuable to the people who are paying for it? If it isn't valuable to your customer, does "early" matter? Does "continuous" matter? Who are you satisfying? You as the leader feel pressure to deliver, pressure not to exceed your budget, pressure to keep your team focused and inspired and working well together. You as the leader must manage these pressures against the realities of time, requirements, and technical capabilities. As Mike shared, "Leaders may pressure technology staff to hit unrealistic timelines or use specific technologies. That pressure can lead to delivery delays and poorly designed products if left unaddressed."

Despite all this pressure—and the pressure is real, no question about that—in the face of all this pressure, look back to the Principle and distill it to this: Your highest priority is delivering value. If you've hit the schedule but failed to satisfy your customer, or if you've come in under budget but failed to satisfy your customer, or kept your team intact and focused and committed but failed to satisfy your customer, ask yourself, what have you accomplished? As Mike shared with me, for Principle 1, the focus is on *value*. "The 'v' in MVP is 'Value.' And this 'V' often gets lost as pressures to deliver increase."

How do you ensure you are delivering valuable software? You start by understanding what valuable software is. And how do you determine what valuable software is? You bring your customer and your agile software development team together. This is the conjunctive aspect of your role. You then cause dialogue through questions. This is the processual aspect of your role. And what comes of the dialogue? Definitions. Clarity. Common understanding. This is the generative aspect of your role. You ask your customer—the end customer or your business owner—to define "valuable software" for you. You ask, and then you listen.

Why ask what matters to your customer?

Consider:

From Alex King, Senior Software Engineer at Amazon Web Services: "As the leader, you ask this question to get clear. You as the leader need to be confident that you are solving the right problem. Invest in deeply understanding the customer problem. When the customer says, 'I want 'X', 'X' may not actually be the solution. Dig in further to understand what attributes of 'X' are appealing to your customer. The best solution may actually be 'Y', that happens to do some things like 'X'. Get a clear understanding of your customer needs and ensure that your team is fully aware of *what* problem you are trying to solve. And encourage, even *require*, that your team verify their assumptions with their customer throughout the development lifecycle."

How do you verify the customer's assumptions? You ask, and then you listen. The process of determining value is in the dialogue with your customer. Mike Brendzal said, "This starts with listening to customers, reflecting back that understanding, and then incorporating that feedback into solutions." We'll get into more detail on this in Principle 4, "Businesspeople and developers work together daily throughout the project." Ask, and listen.

How Do You Think About Satisfying Your Customer?

Consider:

How do you ensure you're aligned with your customer on exactly what they mean when they envision "valuable software"? First, you align with your customer by bringing them into the dialogue with your development team. You as the leader are charged with making this dialogue happen. This dialogue must be more than you and your team asking questions and then jotting down the answers. You and your agile software development team are not order-takers. You are builders. This requires that the dialogue be generative, that it advances you and your team from the current state and that it leads *somewhere*. You are constructors. You *create*.

Here's how a leader at a Silicon Valley company describes his approach and his process for ensuring not only that his team has this dialogue with their business owners. Here's what he does to ensure that the two teams connect and that the dialogue is generative.

Consider:

"The leader needs to be a facilitator. The leader needs to make sure that the teams are prepared to perform at their best and that they're not running into issues. If they are running into issues, the leader helps them overcome those challenges, whether that's a technical challenge or a conflict within the team or anything else. In that sense, the leader is a facilitator. But the leader also raises the bar for the team to make sure that they make progress and perform at their potential."

You as the leader need to ask to determine, what is the source of the challenge? What is the source of the issue? Is it a technical limitation? Is it lack of clarity in the customer's requirement? Is it a changing requirement? Or perhaps it's a conflict within the team. Ask. Listen. Seek to understand. Get clear. Decide.

What makes a leader effective at facilitating that kind of dialogue? How do you as the leader ensure that your team is creating value? Usha Arora from Oracle:

Consider:

"Coming from my experience, I think there are three things. The one thing that I feel makes the leader most effective is making sure that there is transparency and understanding of objectives and goals. What we are trying to do, what we want to achieve when we want to achieve it. That's first.

"The second thing that I think is important is to create a culture and an environment where people are able to make suggestions and provide input without concerns or fear."

"And the third thing that I feel is very important is to measure and provide the visibility on how the team is doing at each stage of the project. This way, the team is able to assess their progress and if necessary, change direction or speed up or make adjustments to make sure that they are able to achieve the goal."

This assessment of the team's performance happens real-time. You as the leader are responsible for ensuring the assessment takes place, and then for ensuring that your team adjusts based on the assessment. If your team is not hitting its commitments for timely delivery and is missing dates and missing deliverable, you as the leader help the team determine where the opportunities lie for making up time.

Consider:

The IT team at TDS Telecom is charged with developing and delivering valuable software and technical solutions in order to transform and optimize its business, the customer experience and ultimately help give TDS a competitive advantage in the markets it operates. Karl Betz, Vice President of Information Technologies at TDS Telecom, shares what his team does: "High frequency releases and higher quality are supported by automated deploys, QA checks, and testing."

How Do You Think About Valuable Software?

What is valuable software from the business owner's point of view?

Consider the perspective of Hamdy Farid, Vice President Business Applications at Nokia. Hamdy is a business owner, and like Mike Brendzal, has deep technical understanding.

Consider:

"Valuable software can be iterative, but it needs to have minimum viable content. Don't give me an iteration or a phase where you deliver the software—'thank you very much'—but it doesn't do anything for me. I have to wait for something valuable. There is a threshold after which the viable becomes valuable to me as the customer. So, what is the minimum viable working product that you can deliver and then you do it in iterations. Even the first release needs to have a minimum size that will get me benefit."

How Do You Think About the Inevitable Mistakes and Failures That Will Happen as You Build Valuable Software for Your Customer?

You as the leader have made clear to your team that its highest priority is to satisfy the customer through early and continuous delivery of valuable software. You've joined your team with the business owner and through the process of dialogue have generated clarity and direction. You've asked your customer what "value" means to them and what "valuable software" is to them. You've checked your assumptions early and often in dialogue with the business owner. You've resolved conflicts and you've cleared misunderstandings. You've listened, confirmed, verified, checked and double-checked, and delivered. You meet with your customer to review your team's latest delivery. You're confident. You're proud. You've moved fast. You have momentum. You meet with your customer, and she tells you you've failed. Now what?

On Risk and Failures

Let's start with risk. If you want to deliver early, you have to change how you're doing what you're doing. And you can only change how you work by introducing risk and instability into the system. You must destabilize the system to change the system. But when you destabilize the system, you risk failure. Doug Lowell has formed his leadership views in a career spanning more than 30 years across high tech, consulting, and healthcare industries. Today, he serves as Vice President, Real Estate and Workplace Resources at Centene Corporation. He shared his perspective based on his years of experience.

Consider:

"You have to figure out how to examine failures. Did the product have an unexpected outage? Or did your customer escalate an issue where the software wasn't working correctly? You as the leader need a robust process around how you examine failure in your system. Do you punish people who failed? How do you get the system to be better at identifying and correcting failures earlier? My premise is that you can't lead by trying to minimize risk because if you do, you'll also minimize productivity and creativity."

At best, you'll get only what you specifically asked for, and nothing more. When your goal is to minimize risk, or if you as the leader declare that you cannot accommodate risk or if your behaviors demonstrate that you will not accept risk because the schedule can't accommodate risk, then you are constraining the team's creativity and putting a boundary around what they can do. Doug shared the dialogue he has with his teams, and he refers to the story of the runner Roger Bannister breaking the four-minute mile to illustrate his point:

Consider:

"Get into the habit of dialogue with your team where you say, 'Wow, that sounds impossible. It sounds like it can't be done. But if it could be done, what would it look like? How could this get even partly done? What obstacles would we have to remove?' That's a way of getting a team to aim really high and to do so in a way that they break down the problem into small enough pieces that they really do start solving it in creative ways.

"There's the famous story about Roger Bannister breaking the four-minute mile. The view at the time—this was up to the early 1950s—was that no one could possibly run one mile in less than four minutes. Four minutes was believed to be a physical barrier of what humans were capable of *mechanically*. The human body just couldn't do it. That belief was held by coaches, by athletes, by everybody. Runners were approaching four minutes but could never crack it, and that lasted for many years. The fastest miles were between 4:20 and 4:05 for years. And then when Roger Bannister broke the barrier in 1954, less than two months later another runner completed a mile in under four minutes. And then another runner did it, and another, and another. So, what does that tell you about the reality of physical constraints versus psychological constraints? We have those all around us in all aspects of life. There are these beliefs about barriers being hard and fast that are not. And so, as a leader, that part of the asking questions is, are these things that we're talking about as constraints true constraints, or are we just failing to find a way around them? Are we failing to see the sub-four-minute mile opportunity in this problem? To me that's a mold-breaking reality to think about. Those exist all over the place."

"In high tech: Moore's law. Microchips will become five twice as fast for half the price once a year. Well, why is it 2X instead of 10X? Why? And another example: a company I used to work for was designing microchips. Chip design has gotten into a microscopic physics state where you're talking about widths of molecules that are on a chip. Widths had been reduced to six nanometers. The next target was four nanometers. In terms of dimensions and capacity of a chip, even the customers and the chip making business started asking companies like mine that were working on the design process, 'When are you going to have a program that will help me break the four-nanometer barrier?' It became an incremental problem to solve instead of being a barrier. Now everybody's looking at the next increment. The thinking and the problem-solving became structured around incremental thinking instead of trying to break through what seemed like an impossible barrier. All the competitors had been playing the same game. Nobody was saying, 'How do we rethink this all together?'"

As there is a generative aspect to thinking about risk and taking risks, there is a generative aspect to failures. Shaye Robeson leads the enterprise and solutions architecture team at UScellular.

Consider:

"Failure is part of the process. I don't think that we expect failure often enough and that we tell people that we expect failure. The more uncertainty you have, the more you need to iterate, and the more you need to iterate, the more you need to expect that failure is part of the process. You want to find the failure as quickly as possible and as early as possible so that you can do the next iteration and move on. Failures are part of the process. This is how it works. This is how we deliver quickly. It's because we iterate and quickly fix issues, instead of trying to be perfect and then testing and testing and testing until we think it's perfect. Then after all the testing, if something doesn't work, we get yelled at and people are disappointed because we took all this time and created this false expectation that when we were 'done,' and that the finished product, the deliverable, would be perfect. And then we deploy, and we find out it's not perfect. It never is."

The reality of periodic failures must be understood and accepted not only by you as the leader but also by your team and by your business owner. Be clear up front, early in the dialogue, that failures are expected, and they will occur. Failures are generative opportunities. Shaye sees it this way:

Consider:

"Failure needs to be contextualized. It needs to be understood and accepted as part of the process. Too often failure is defined as something that either does or doesn't happen until the end. Leaders need to communicate that failure is part of the process and that it's expected. And then when they see failures, and they inevitably will, they need to talk about it. They need to tell stories about failures. They need to demonstrate through their actions and their narratives that failures are expected and are as much a part of the process as anything else. And that when the team fails, they adjust. And that this is exactly what we all should be doing. Our goal is to generate that failure as early as possible so we can get to the next iteration. That's considered part of the process."

Are You Ready?

You started by deconstructing this Principle to get at its meaning. You have defined the words and you understand the words. You've defined your objective with this Principle by starting with the end in mind. You've defined what matters most. You know why this matters most because you've asked great questions of your team, your boss, and your customers. You've asked great questions and listened, really *listened*, to what you were told, and you have learned from their stories.

Now that you are clear on what it means to satisfy your customer through early and continuous delivery of valuable software, what do you do? You are the leader. It is time to act.

Do

You as the leader must bring your customer or your business owner together with your agile software development team. In the dialogue, you and your team help the customer define what "valuable software" is. What a customer wants and what a customer needs often are two different things; what the customer needs matters more than what they want. It's the difference between "nice to have" and "necessary to have." You as the leader must help your team get there, and you do this through the process of dialogue. The same notion applies to software developers and the product they create. Developers too can struggle between what they want to develop and deliver, and what they need to develop and deliver. You as the leader must lead your team to help get the customer clear on the distinction. Accoutrements are not requirements.

Key Takeaways

1. Listen.
2. Bringing together your customer and your agile software development team is a necessary part of the development process.
3. The dialogue between your team and your customer leads to clarity on what "value" means to your customer.
4. The process of dialogue must generate outcomes that are clear, concise, specific, and valuable.
5. Change does not happen if you do not take risks.
6. Failures are a necessary part of the process.

Reference

Denning, S. (2018). *The age of agile: How smart companies are transforming the way work gets done*. American Management Association.

Agile Principle 2: "Welcome Changing Requirements, Even Late in Development. Agile Processes Harness Change for the Customer's Competitive Advantage"

Abstract

Delivering valuable software frequently is the name of the game. Requirements will change throughout the lifecycle of your project. Some changes you will be able to accommodate sooner and faster than other changes. Requirements can change for several reasons: new information, clearer understanding, changing market dynamics. Your development process must not only allow for change, but encourage and generate change, so that the product you deliver is valuable to your customer at the time you deliver it. Customers require what is valuable; customers don't require what was valuable last year.

Vignettes from leaders at Centene Corporation, Oracle, Amazon, and Nokia illustrate the objective, the challenges, and solutions relative to this principle.

Leaders will experience changing requirements throughout the development process. To address these challenges, leaders establish a repeatable, sustainable process for generative dialogue with their customer. The leader ensures roles and responsibilities are clearly defined. The joint team—developers, business owners, testers—executes in an agile manner, which in turn allows for, accommodates, and even encourages changing requirements throughout the process. This iterative and conjunctive process generates products and solutions that enable the customer's competitive advantage.

What Does This Principle Mean?

What does this Principle mean? Here's a way to think about this: "We welcome change. We work in such a way that change does not disrupt us or slow us down. We change when our customer needs us to. We join what the customer says they want and need with what we believe we can create and deliver. It's not about being right or wrong. It's about together generating a better outcome."

******

Let's deconstruct this Principle.

"Welcome changing requirements." Listen for the difference between lack of clarity and changing requirements. Listen for what is changing and why it is changing. Listen for clarity and make the distinction between changing requirements and changing priorities. Are requirements changing because the customer's wants and needs have changed? Is the competitive marketplace changing? Was the requirement simply missed? The answer matters because it could point to an opportunity for process improvement the next time around.

"Even late in development." Because your team is developing and delivering in sprints, "late" can be more easily managed than "late" in waterfall delivery. "Late" might be late for one sprint only.

"Agile processes harness change." Jeff Mander, Director of DevSecOps for Digital Platforms at UScellular, shares this approach for measuring change: "Scope change is measuring any changes that occur during the active two-week sprint on scope that was already committed in the sprint planning meeting. This can be an increase or decrease based on the item. When there are changes in scope that we include or exclude, it is also estimated and tracked via story points to be consistent. Ideally, your development capacity includes wiggle room for change or for accommodating something that is more complex than expected. We do have items that we know we can pull forward into a sprint if things are removed from scope or we have unplanned extra capacity."

"For the customer's competitive advantage." Your agile software development team exists to deliver value to your customer. Helping create competitive advantage is your team's raison d'etre.

You Are the Leader. What Do You Do?

Start with the End in Mind

What will this look like when you've achieved it? When this Principle is in place, your team will be focused not only on developing the solution your customer needs, but will also be monitoring the marketplace, watching and listening to what your customer says and what other customers are saying. You will be making the changes and adjustments that best serve your customer rather than being constrained by the choices and decisions your customer made when the marketplace and the competitive environment were different. You are building and delivering for today's needs, not yesterday's requirements. And this is the way your team will measure its progress and report out to its customer: meaningful progress toward what they need now, not what they needed yesterday.

This should be common sense. What are you going to do, deliver what the customer doesn't want? Deliver what the customer can't use? Deliver what the customer says will not work for them? Choose to dissatisfy your customer? Your customer has no clearer or more accurate view into the future than you do. Neither of you has a crystal ball. Neither of you can predict what tomorrow's competitive marketplace will bring. So, it's unrealistic and, frankly, unfair to hold your customer accountable to requirements she defined a year ago that she believed would be exactly what she needs a month from now. Her view should be a realistic view, but it won't be perfect. She doesn't have perfect knowledge of the future and neither do you.

Listen and Learn from Others

How do you get to the "end in mind?" You start by asking great questions, and then you listen.

- How is the customer going to use the product?
- What's the source of changing requirements?
- Why might requirements change?
- What is your customer's expectation when requirements change?
- Why should your team accommodate changing requirements?
- How can I help?

Listen to your team.

Your team needs to know what they're being asked to do. Simple enough. They also need to know *why*. When they understand *why*, they can bring to bear their experience with similar challenges. They may find ways to solve for today's challenge in novel and innovative ways. Consider the insight from Muhannad Obeidat, Vice President, Software Development, at Oracle, on the importance of clarity in the requirements and clarity in the *why* that is driving the requirements.

Consider:

"Clarity of purpose. What is the purpose of what you're doing? I've been using a concept from Simon Sinek's book, *Start With Why* (Sinek, 2011). I use that concept at the beginning of every project, from Release 1. We always say before you tell anybody what the feature is, tell them why they need something. And then you tell them what it is and how to use it. We tend to immediately say here is what the feature is, and here is how to use it. But why does it exist? That's way more important than what it is and how it works. And you realize that it's a natural thing to do, but you don't always do it unless you take a step back and say to yourself, 'I should always start with why.'"

Muhannad's insight and experience underscore the importance of clarity of purpose. And clarity of purpose is achieved in the process of ongoing dialogue with your customer. Clarity of purpose and clarity in the requirements does not imply that you need to tell your team exactly what to do and exactly how to do it. Quite the opposite. Consider Doug Lowell's description of the importance of clarity.

Consider:

"Teams like clarity of direction. They don't necessarily like a lot of direction, but they like it to be clear on where they're trying to get to. I've seen teams that don't have that, and they get frustrated by that because they feel like they're hunting for a direction. It creates a lot of rework. People go down one path and then learn that that's not really what was wanted, and they've wasted a lot of time. There's nothing that frustrates the team more than feeling like they're putting their heart and soul into something and then learning that that's not what was wanted in the first place, and that the objective could have been more clearly established at the beginning."

Alex King, Senior Software Engineer at Amazon Web Services, emphasizes this point. Leaders must ensure "that the team has a clear long-term goal for what outcome they want to achieve. Not necessarily the specific product, but a good pulse on what problem they are solving."

How do you do this? Through dialogue with your business owner. Your business owner is working on behalf of her customer, to establish a clear vision of what matters most to her customer. The process of dialogue with your business owner will distill the customer's vision, and that process will generate clarity of what the team needs to do and why they need to do it.

One of your responsibilities as the leader, and one of the roles of your team, is to identify changes as early as possible and discuss them openly and widely. Remember, changing requirements are different than changing priorities. Priorities that change midstream hit your team much harder.

But customers do change their minds. Business owners change their minds. The world around us changes. So, for you as the leader, how do you accommodate a true change in requirements versus lack of clarity in the requirements?

You must be able to accommodate a true change in scope, and you have to be able to communicate these changes to your team in ways that provide clarity and help create momentum. Teams can feel deflated, frustrated, even angry when they are asked to discard what they've been working on. This is seen in practice, and it's also described in the academic literature on this topic. "Simpson and French (2006) draw on Bion's work to highlight the value of 'negative capability' for effective leadership practice. They argue that this capability embodies patience and the ability to tolerate uncertainty, frustration, and anxiety in ways that enhance leaders' ability to think in the present moment" (Collinson, 2014). You are the leader. Your ability to be effective in the face of change, and to communicate these changes in an honest, direct, clear, and optimistic way is essential for keeping your team focused on solutioning.

You as the leader don't need to know exactly how to design and code for the change. Bring your team into the dialogue. In Doug Lowell's view, "Leaders don't have to establish this clarity on their own. They can do brainstorming with the team to establish the direction the same way that they would any other creative process. Because even if you know the outcome you want, there may be any number of ways of getting there."

You are the leader of an agile software development team, and you know this. You know and your team knows that likely there are multiple ways to create and develop solutions. As you and your team develop solutions in a changing landscape, Alex King from AWS encourages leaders and development teams to "make as many two-way decisions as possible. Prioritize flexibility in design and development. If you can keep something generic without loss, then do it."

There's a critical distinction that bears calling out here, and that's the distinction between requirements and priorities. Principle 2 is clear: the Agile Software Development team (the ASD) welcomes changing requirements, even late in development. Why might requirements change late in the game? Is it indecisiveness on the part of the product owner? Is it less a change of requirements than greater clarity in the requirements? Or is it changing market conditions that cause the product owner to alter her requirements late in the process? These questions need to be answered in the spirit of continuous improvement. To be clear, there's a very important difference between changing requirements and changing priorities. When priorities change, work streams stop. Projects stop. The pace of development is disrupted. And your developers can be left feeling like management can't make up its collective mind, or that leadership is not aligned. Worse, they can be left feeling like they've wasted a lot of time and effort. Agility is a strength; indecisiveness is a weakness. This Principle is connected to Principle 8, "Agile processes promote sustainable development. The sponsors, developers, and users should be able to maintain a constant pace indefinitely." True enough, a change in priorities does not by definition disrupt the *process* of agile software development. But a change in priorities *does* disrupt the *content* of what the team is developing. And when the content changes significantly, work streams are disrupted, trains of thought and development are derailed, development that has been completed to date becomes marginalized or, in the worst case, the development is wasted because it is no longer needed. Few things demoralize a development to a greater degree than time, energy, and development that is thrown away because management couldn't make up its mind.

Listen to your customer.

You've got to put in the work to understand what your customer wants, and then you go build it. You must be prepared for the customer's wants and needs evolving as they continue to learn more about their own customers. This level of understanding requires that you have a generative relationship with your customer. The generative aspect of your connection is this: you listen, you seek to understand, you offer, probe, question, challenge, and in the end, co-create a vision for what your team will deliver. The dialogue with your business owner is two-way and is aimed at generating a solution.

Nothing frustrates a customer or a business owner more than a finished product that misses the mark. Customers and business owners can live with defects. They can live with imperfections. What they cannot tolerate, and should not be expected to tolerate, is a software development team that does not listen to their requirements and that doesn't understand the reality of the marketplace. It is the job of the leader—*you*—to stay connected with your customer.

Are You Ready?

You started by deconstructing this Principle to get at its meaning. You have defined the words and you understand the words. You've defined your objective with this Principle by starting with the end in mind. You've defined what matters most. You know why this matters most because you've asked great questions of your team, your boss, and your customers. You've asked great questions and listened, really *listened*, to what you were told, and you have learned from their stories.

Now that you are clear on what it means to welcome changing requirements, even late in development, and what to harness agile processes for the customer's competitive advantage, what do you do? You are the leader. It is time to act.

Do

Clarity matters. Get clear in your own mind, then get clear with your team and your stakeholders, on how the agile development process works, and why it works this way.

Be Clear on How the Agile Development Process Works Share your process with your stakeholders. It is important that the business owners understand why you work in the manner that you work. Be clear. It is valuable for them to understand this so that they can work with you. Their role is not to conform to the way you work, but they should have a basic understanding of your approach.

Next, listen to your customer.
Next, ask your customer again, and listen again.
Next, hold fast to your delivery dates.
Next, deliver. This earns you the credibility that you need to establish with your key stakeholders.

Be Clear on Roles and Responsibilities Establish boundaries. The importance of boundaries becomes especially relevant when the requirements, development, and delivery process is iterative. To be clear, the business owner should not be making design decisions. They should be defining requirements that enable and support what matters to the customer. *Your* team makes the design decisions. Establish this boundary early and clearly and revisit it often to ensure that the business owner and your team play their roles and deliver on their core responsibilities.

Your role at this stage is to establish conditions that foster growth. You do this by connecting your team with the business owner. This connection does *not* replace *your* connection with the business owner. Instead, this connection heightens your team's sensitivity to the business owner and deepens their understanding of what the business owner needs. When you've established this connection—your team's relationship with the business owner and with the product—you've established conditions that generate creation and growth.

Key Takeaways

1. Listen.
2. Requirements change. This is a fact. Spend time in dialogue with your business owner understanding why the requirements are changing and what is now required.
3. Clarity of what the customer needs, why they need it, and how they will use the product is paramount.
4. Clarity is achieved through the process of dialogue.
5. The purpose of clarity is to generate understanding and to enable new thinking and new solutions.

References

Collinson, D. (2014). Dichotomies, dialectics and dilemmas: New directions for critical leadership studies. *Leadership, 10*, 36–55.

Sinek, S. (2011). *Start with why: How great leaders inspire everyone to take action*. Penguin.

Simpson, P., & French, R. (2006). Negative capability and the capacity to think in the present moment: Some implications for leadership practice. *Leadership*, 2(2), 245–255.

Agile Principle 3: "Deliver Working Software Frequently, from a Couple of Weeks to a Couple of Months, with a Preference to the Shorter Timescale"

Abstract

Delivering valuable software frequently is the name of the game. Your customer wants and needs solutions as quickly as you can deliver them. Solutions delivered late are less valuable than solutions delivered early. Customers want and need you to create valuable opportunities for them today, not tomorrow. Customers want and need you to solve their costly problems today, not tomorrow.

Vignettes from leaders at Oracle, Amazon, and Nokia illustrate the objective, the challenges, and solutions relative to this principle.

The most effective way to satisfy customers with valuable software sooner rather than later is to bring them along on the journey through the process of software demonstrations, or demos. Demos provide the opportunity for developers and customers to learn together and to co-create. Honest feedback that serves the objective of the project is imperative to progress. Acknowledging all feedback is imperative to maintaining trust. Honest and constructive feedback that is addressed by the development team generates progress toward the goal of delivering working software frequently.

What Does This Principle Mean?

What does this Principle mean? Here's a way to think about this: "We like to move fast. We prefer to move even faster. And our customers prefer even faster than that."

Let's deconstruct this Principle.

"Deliver." What do we mean when we say "deliver?" This means exactly what it says: your team puts their work product in front of their internal stakeholders and shows them what the work product can do. No more, no less.

"Working software." What do we mean when we say, "working software?" Simply put, this is software that functions, that *does* something.

"Frequently." Here, the Principle gets interesting. "Frequently" means different things to different people. For our purposes, "frequently" will be defined as delivery between 2 weeks and 2 months. Most teams—from developers to business owners to executive sponsors to customers—will prefer the shorter timeframe.

You Are the Leader. What Do You Do?

Start with the End in Mind

What will this look like when you've achieved it? When this Principle is in place, your team will be delivering smaller, working components of the product, and will be delivering, or dropping, those components frequently. Your team will not be waiting to ship the product until they consider the product done—all requirements met, development done, testing done, and delivery completed. Instead, your team will be delivering what they get done, when they get it done, on a schedule that you've agreed to with your business owner. Delivery can take place in one of two ways: either in scheduled and published in 2- to 4-week cycles, or it can take place continuously, in a Continuous Integration/Continuous Delivery model (CI/CD). This is what your customer expects because this is what you negotiated and agreed to in the process of dialogue with your business owner, and so this is what you will deliver. Whichever way you've agreed to deliver is fine, so long as *you and your business owner agree*. Whichever way your team has decided to deliver, what matters most is that you and your team do what you said you will do. Stanford Professor Emeritus Ron Howard captures this notion: "Say what you mean and do what you say." Be accountable and be reliable.

Listen and Learn from Others

How do you get to the "end in mind?" You start by asking great questions, and then you listen.

- Ask your team: How frequently can we deliver?
- Ask your team: What do we need from our business owner so that we can deliver frequently?
- Ask your business owner: How frequently would you like us to deliver?
- How can I help?

We have deconstructed this Principle to get at its meaning. We have clearly defined the words. We have defined the objective with this Principle by starting with the end in mind. We have asked great questions and listened, really *listened*, to what we were told. Now it's time to learn from others.

First, get clear on requirements. We discussed this above with Principle 1.

Second, establish and commit dates with your business owner.

Third, demonstrate your progress to your fellow agile software developers and, most importantly, to your business owner. This event is the demo, or demonstration. What you're doing with the demo is more than sharing progress. Your team is doing more than showing integrations. Demos, at the foundation, are about trust. Mike Brendzal emphasizes that "the theme of all this is about the importance of building trust through actions. Initial alignment needs to turn into alignment at the detailed level and backed up through actions and decisions."

What is a demo, and why do you do it?

What is a demo? An integral part of the process of delivering working software regularly and frequently is the demo, or the demonstration, of the development team's progress to date. Demos are performed to elicit feedback from your business owner and from other developers. From Scaled Agile Framework: "The System Demo is a significant event that provides an integrated view of new Features for the most recent iteration delivered by all the teams in the Agile Release Train (ART). Each demo gives ART stakeholders an objective measure of progress during a Program Increment (PI)" (Scaled Agile, 2021). It's the method for assessing the solution's current state and gathering immediate feedback from Business Owners, sponsors, stakeholders, and customers. The demo is the one real measure of value, velocity, and progress of the fully integrated work across all the teams." The demo is the opportunity for "business owners, executive sponsors, other Agile Teams, development management, customers (and their proxies) [to] provide input on fitness for purpose for the solution under development. The feedback is critical, as only they can give the guidance the ART needs to stay on course or make adjustments" (Scaled Agile, 2021). Another excellent description of the purpose of the demo: "A Sprint Review (Demo) provides the platform for the Scrum Team to showcase what they accomplished during the sprint while creating the opportunity for key stakeholders to inspect the increment and adapt the Product Backlog, if necessary."

Consider:

"Before we even go out to the consuming teams, we have a concept called Demo Day. Before we take our product public, we do Demo Days with the teams every two weeks to four weeks. Developers that own different areas come and demo what their feature looks like right now and how it's going to be used. And then they demo that, first within the smaller group of the product team that I'm responsible for, then we demo it in front of 60+ user across the organization that consume our product" (Interview with Muhannad Obeidat).

Consider:

"Monthly demos [at Amazon] were a very open forum. If you wanted to share something, you just signed up. If you wanted to see what other teams were doing, you went to the demo. We went around the horn and asked each developer to show us what they built. It could be anything. It didn't have to be working code. Someone might show up to the demo and say, 'I just completed this design, and this is what was challenging about the design. We came up with this new methodology to scale this particular type of service, which has been a problem in the past. I want to show you.' Sometimes it was a demo on an operational problem that got fixed. It wasn't necessarily always, 'Hey, look at this new functionality for the customer,' or, 'I solved a problem,' or, 'We minimized or eliminated a distraction,' or, 'I improved a process or a procedure that had been very time-consuming,' or, 'I automated a solution.' It could be test cases that got automated, things like that. It all depended on what people wanted to share. Demos could be showcases or conversation starters or requests for help, whatever people wanted to share. A lot of times these were learning opportunities too, where senior level folks would share strategic information about a design, and why X design was better than Y design.

"The demos would generate energy and excitement in the room. Developers and business owners would see the next iteration of the software, and they would say, 'Ohh, I remember when this only did *this*, and now it does *that*, and oh, that's very cool!' There was a lot of pride and ownership in the room on Demo Day. I could feel the momentum building, the momentum for getting the next thing out the door" (Interview with Heather Ackenhusen).

Muhannad Obeidat describes the importance of feedback in the demo progress. Consider:

"Demo feedback is critical. Get it very early and continuously. Include ETQA in the audience for demos so that they can start getting an understanding of what's coming their way and they can start planning for it. In demos, agile software development teams need to connect with their users, versus expecting customers to connect with them. Show the customer what is being done, not everything that can be done, and certainly not how, technically, it gets done. Stay focused and speak the customer's language."

With the demo, you are creating a condition that fosters growth. This is a great example of generative leadership. You are bringing together your agile software development team with the business owners in a forum to see and review and question and challenge and advance the work in progress. This may feel chaotic at times—developers and business owners together facing and addressing uncertainty—but it is necessary for moving quickly in the right direction. You're showing what's working.

Recall Heather Ackenhusen's experience at Amazon. Demos were showcases and learning sessions. This is the very definition of "generative." Creative, provocative. Exchanges and interchanges. Giving and taking, pushing, building, extending, encouraging, inspiring. Doing this together, on purpose for a purpose. The team is a true team: different roles, different positions, all on the same side and all for the same meaningful and valuable cause, a cause that smart people on a mission can rally round and work hard to support.

Remember, the fundamental purpose of the demo is to elicit feedback from your business owner. Feedback is necessary. Different points of view are necessary. Feedback, delivered constructively, can generate new ideas, changes, tweaks, improvements, and innovation. But for the feedback to be valuable, the feedback must be *heard,* and it must be *addressed.* Researchers Michael J. Arena and Mary Uhl-Bien summarized a body of research and concluded that "findings suggest that what is needed in complex organizations is an adaptive response—one that involves engaging, rather than suppressing, the tension generated in the conflicting perspectives of the operational and entrepreneurial systems" (Arena & Uhl-Bien, 2016).

The feedback isn't always about fixing something or changing something. Feedback loops in Heather's experience sometimes served to provide positive reinforcement. In some cases, developers or product owners would say, "The customer usage of this new feature went up by X percent, or X number of people are now using this feature. 'You, developer, delivered this new feature, and here's the positive impact it's having on our customers.' Teams would want to iterate on this because they were getting the signal, the feedback, that customers liked it. There was a constant sharing of progress and results and feedback, and it created a virtuous cycle."

Are You Ready?

You started by deconstructing this Principle to get at its meaning. You have defined the words, and you understand the words. You've defined your objective with this Principle by starting with the end in mind. You've defined what matters most. You know why this matters most because you've asked great questions of your team, your boss, and your customers. You've asked great questions and listened, really *listened*, to what you were told, and you have learned from their stories.

Now that you are clear on what it means to develop working software frequently, from a couple of weeks to a couple of months, with a preference to the shorter timescale, what do you do? You are the leader. It is time to act.

Do

What makes for a great interaction in a demo? How do you do this, and do it well? It's easier said than done, for sure. Demos can be a source of pride and excitement and momentum, or they can discourage and even alienate business owners. In Muhannad's experience, interactive demos sometimes can be a challenge. Sometimes, it's difficult "to get people to speak up, to ask questions, to give us real feedback." Mike Brendzal has experienced this; so has Hamdy Farid. So how do you conduct a successful demo that generates progress? Mike encourages leaders and their development teams not to give up: "Demo early and frequently, prioritize the feedback, even when it creates more work. Clearly show how the feedback is tracked and addressed." If you're going to take the time and make the effort to conduct demos, do your best to cause great interactions in the demo. Hamdy shares his experience:

Consider:

"You must have emotional intelligence both as the business owner giving the feedback and as the developer receiving the feedback. As the business owner, it's critical that you give the feedback in a way that isn't threatening or confrontational. Here's why: if our customer has a major problem with the software that we've delivered to them, and I'm talking about a really significant, major problem, and we've had everybody working 24/7 for the last five days trying to sort it out, and then I come into the room and start telling them what they should have done differently—'You should have done this, you shouldn't have done that'—well, this doesn't help. This doesn't help you. It doesn't help them. It doesn't help anyone. So, you hold on to this comment, and instead of saying it to the entire team, you talk with the leader, you talk to a select few. You still need to provide your perspective, but you've got to do it the right way. When I handle it this way, I find that they take the feedback because they want to make the product better. Remember, engineers and software developers like to build stuff. And if the feedback makes the creation better, why not give them feedback? If I'm the developer or the builder, I will take that feedback and improve my product because my product is reflection of our company."

Mike sees demos as a way to show what has been or is being built, to collect feedback on what was demonstrated, and to collect feedback on what might be built in subsequent sprints. As the leader of the development team, you have a role to play in helping create a positive and generative demo session.

What impedes demos from being productive? What can derail a demo? One leader, who asked that I not identify her, shared her and her team's experience with unsuccessful demos:

1. Demos with 100+ attendees.
2. Feedback not tracked, not prioritized.
3. No test environment or screenshots for users to evaluate what was built against the context of their needs.
4. Demos led by developers who have little understanding of the users' needs, roles, or terminology.
5. Answers to participant questions that are incorrect or designed to appease rather than address the feedback.
6. No testing of anything in the demo.
7. No time for providing feedback.
8. No process for formal collection of feedback.
9. No process for assessing feedback.
10. No accountability for the disposition of the feedback.

Today's game is about speed. The marketplace is changing, and the competition is moving fast. You as the leader need to ensure that your team is moving fast, too. One of the things that frustrates business owners and product owners alike is the sense that your team is not moving fast enough, that your team is not enabling the business to keep pace with the competition. But how fast do you need to be? You must be fast, but not so fast that you miss the mark and fail to deliver what the business needs from you. You need to be fast, but not as fast as you *can* be, rather be as fast as you *need* to be. And how fast to you need to be? Faster than the person or company you are competing against.

You are the leader. You and your team own creating the environment that joins your developers with their stakeholders. You own facilitating a process that is repeatable and sustainable, and that elicits generative feedback.

Key Takeaways

1. Listen.
2. Your customer wants and needs you to deliver working software frequently.
3. Elicit feedback in the demos.
4. Demos combine all the dimensions of today's leadership.
   4.1 They are conjunctive in bringing together different teams across the organization.
   4.2 They are processual in that working software is being demonstrated and reviewed.
   4.3 They are generative based on the input and the feedback that is provided to the developers.
5. Act on feedback. This doesn't mean that you must make changes based on the feedback. It does mean that you must acknowledge all feedback you get and share what you decide to do with it.

References

Arena, M., & Uhl-Bien, M. (2016). Complexity leadership theory: Shifting from human capital to social capital. *People and Strategy, 39*, 22–27.

Scaled Agile. (2021, March). *Scaled agile framework*. Retrieved from www.scaledagileframework.com; www.scaledagileframework.com

Agile Principle 4: "Businesspeople and Developers Must Work Together Daily Throughout the Project"

Abstract

Delivering valuable software frequently is the name of the game. Businesspeople and developers must be aligned on the objectives of the project. And they must be aligned on the deliverables due at each stage of the project. Priorities and deliverables can change, and this makes daily connection and daily alignment that much more important. When businesspeople and developers work together daily—when they commit to generative dialogue and meaningful progress—they create the products and solutions that matter most to their customer.

Vignettes from leaders at Oracle, UScellular, and Amazon illustrate these leadership challenges.

To ensure the development team makes progress every day, you as the leader must connect the businesspeople with your developers. Your team must commit to generative dialogue. This dialogue will help ensure that your team gains insight from the businesspeople. The questions your team asks in this dialogue help achieve clarity. The team reaches alignment. The teams jointly prioritize and reprioritize as customer needs evolve and as the marketplace evolves.

With insight, clarity, alignment, and prioritization, the team is positioned to make meaningful progress toward delivering valuable software quickly.

What Does This Principle Mean?

What does this Principle mean? Here's a way to think about this: "We talk with and listen to our customer. We do more than work side-by-side. We collaborate in ways that generate clarity. We prioritize, we adjust, and we generate solutions."

Let's deconstruct this Principle.

"Businesspeople." For our purposes, businesspeople are the business owners. Business owners represent the needs and the interests of the customer.

"Developers." This is your agile software development team. Your team includes the product owner. Your development team mustn't be order-takers. Rather, your agile software developers are collaborators and enablers and creators.

"Work together." Working together means more than sitting in the same room. It means more than being a two-dimensional face on a video call. Working together means *communicating*. You can be in the same room, but you must be communicating. You can be a two-dimensional face, but you must be communicating. When you're communicating, you're connecting. When you're connecting, you're generating outcomes that cause progress. Working together includes challenging. Your team must challenge in a manner that generates clarity. They must challenge assumptions, not intending to second-guess or pretend to be expert in a space they're not expert in, but rather in the spirit of creating more useful and more valuable solutions. They challenge requirements to get clear and aligned on how the customer will use the product.

"Daily." Daily means exactly that: you meet and work together and communicate every business day. No exceptions.

You Are the Leader. What Do You Do?

Start with the End in Mind

What will this look like when you've achieved it? When this Principle is in place, your team will be conducting daily standups that include, at a minimum, the software developers, the scrum master, and the product owner. Some teams also include the business owner in the daily standup. These daily meetings will generate insight, clarity, alignment, prioritization, and meaningful progress. The basis for each of these is *interpersonal connection*. More on each of these below.

Listen and Learn from Others

How do you get to the "end in mind?" You start by asking great questions, and then you listen.

- Ask your team and your business owner: What does "working together" look like?
- Ask your team and your business owner: What do we intend to accomplish by working together daily? What is your objective? What is the desired outcome?
- How will your customer use the product?
- "How can I help?"

We have deconstructed this Principle to get at its meaning. We have defined the words and we are clear on what the words mean. We have defined the objective with this Principle by starting with the end in mind. We have asked great questions and listened, really *listened*, to what we were told. Now it's time to learn from others.

Connecting to Collaborate

Connecting requires listening. Connecting requires caring. Find the developers you need and the people from the business that you need, and then put them together. Build a team that is collaborative and connected and feels a sense of togetherness. They don't have to be best friends, but they will be more successful if they are collegial. They don't have to establish a social bond, but they do have to be connected closely enough to understand each other. "*How people interact* matters a great deal for what comes out of the interaction" (p. 52), writes Gisela Backlander. "Previous research on team creativity has suggested that how people relate to and interact with each other is the 'single most important stimulus' for encouraging creativity (Hemlin, 2009)" (Backlander, 2019).

As you seek to understand, you must also seek to be understood. Seeking to understand requires empathy; seeking to be understood requires patience and humility. It doesn't mean that you repeat yourself again and again, or that you get louder and louder. It means that you find ways to connect. You seek to understand how your colleague learns, how they listen, how they express themselves, and what they respond to. This is the basis of connection and a basis of communication.

The connection between the business owner and the product owner can be and, frankly, should be, one of the most generative relationships that your team will build. This relationship is the basis for meaningful dialogue, which in turn is the basis for *gaining insight, realizing clarity, achieving alignment, establishing prioritization*, and *making meaningful progress*.

Gaining Insight

The daily communication with your business owner must elicit her insights. But insights into what? Insights into how priorities are being established and what these priorities are. Insights into what the customer cares about. Insights into how the customer will use the product. Insight into what matters most.

Reaching Clarity

The conversations between your team and the business owner also must elicit clarity of the business owner's vision for the product her customer wants and clarity of how her customer will use the product. This dialogue is not the time or the place for high degrees of specificity in what specific features are desired, or what specific functionality must exist. Instead, it's a conversation to distill general wants and needs into a clear vision of what the customer most wants, and how the customer will use the product.

Achieving Alignment

You are the leader. When you're leading a development team or a development organization, how do you ensure that your development team stays in sync with the product owners? Dinakar Hituvalli, Group Vice President of Product Development at Oracle, has lead development teams for more than 20 years.

Consider:

"We have applied agile methodologies for the last several years now. We have multiple ways of ensuring that our development teams stay in sync with the product owners. We have stand-up meetings three times a week, which include representation from engineering, product management, Quality Assurance, and all aspects of product development.

"In these stand-ups, we go through the backlogs for that sprint. We have the product managers clarify the answers to any open questions. These are quick meetings, usually 15–30 min, truly stand-up style meetings. We go quickly. We ask questions, and we also go through any questions that they have.

"In addition to this, at the beginning of a sprint, we review the design work that's produced by the product managers. We have a design walkthrough where they quickly walk through the designs. In these walkthroughs, engineering and others who are consumers of that functional design attend, and they can ask whatever questions they want.

"We have a very systematic approach. A few days before the end of the Sprint we have a development walkthrough. In these meetings, engineering demos what they've built in the sprint for product managers to make sure that it's as per spec.

"If the product managers think something's off or think that requirements have been misinterpreted, they'll give the feedback. That way, developers still have a few days to correct it within the sprint. Very rarely there are some substantial differences between what the developers built and demoed, and what the product managers expected. In that case, the refined requirements become a backlog for the next sprint."

What happens when there is disagreement or misalignment? What happens when the progress on the product isn't meeting the business owner's expectations?

The answer and the next steps depend in large part on two things: first, the quality of the relationships between your development team and the business owner. Second, it depends on the effectiveness of the process and the quality of the dialogue in the process. Dinakar and his teams don't frequently encounter significant differences between the product manager's requirements and expectations, and what the developers build and demo. Why don't they? "I believe this is because there was a design review at the beginning of the sprint. Given so many touch points and close coordination, developers can be sure that they are building per the spec."

"It also depends on how seasoned the teams are. Fortunately, we have development managers who are quite seasoned. They understand functionally what we're trying to achieve. So, they're able to interpret the spirit and the intent behind the functional designs, and then get the teams to build it out."

Alignment is critically important for teams throughout the organization: at the senior-most levels of leadership and at the front line. When teams are aligned, the energies of the leaders and the energies of the team can be oriented to problem-solving and making progress.

Misalignment.

What happens when teams are *not* aligned? You are a leader. You know that misalignment occurs, and conflicts arise. You are the leader. What do you do? Dinakar shares his experience.

Consider:

"There will be differences of opinion. The fact that some of the development managers or even the development directors are very seasoned, it can be a double-edged sword. They end up challenging the product manager sometimes. They'll say, 'I don't think it should be done this way. I think it should be done in a different way.' Depending on the maturity of the product manager, some of them look at it objectively, and if it's good feedback, they incorporate it."

"Some of the product managers feel challenged, though, and sometimes they don't react to challenges very well. In such cases, it comes to a leader for resolution. A leader will step in and make the final call. But that doesn't happen very often. Given that the teams have been working together for a while, usually they get along well, there's not a lot of conflict.

"Sometimes specs are not clearly laid out. Sometimes they're open to interpretation and maybe a little ambiguous. We convey feedback to the product manager, or we ask that they please review our concerns and address them." The process of dialogue leads to generative outcomes when positive relationships are in place and where the teams are aligned.

Establishing Priorities

Together, the businesspeople and the developers—the team—prioritize what's needed.

The UScellular team charged with developing, delivering, and operating its digital properties operates this way. The business owner prioritizes the defects. The team has daily standups—virtually or in person. The output of the standup is shared with the Operations team to triage and coordinate, then to the Development team to develop and schedule and deliver. Clear boundaries exist between the development team, the operations team, the testing team, and the business owner. These clear boundaries help the team establish clear roles and responsibilities.

Making Meaningful Progress

Sometimes it goes well, sometimes it doesn't. When it goes well, teams are clear on their roles, they provide input, they debate, they disagree, and they commit to progress. You as the leader are leading this process. You are leading change, but you are not *controlling* the change. You as the leader have established the conditions for growth, but you are not controlling what that growth *looks like*.

Progress may at times feel chaotic, particularly when progress doesn't come easily. And at other times, the teams will not be satisfied with the output of the dialogue and the output of the development work. That's okay. Your team is not the first to have done something that didn't turn out to be exactly what the business owner needed. Your team is not the first, and it will not be the last. This happens, and it's okay.

Consider this reflection from a former development lead at Amazon:
Consider:

"If it didn't work out, it's OK. No harm, no foul. There are certain points in time where you do have to cut bait. Here's what sometimes happens: We have a hypothesis, we test it, and then we tweak it and test it one way, and then we test it another way, and it simply is not working. So, we cut bait and let it go. I would say that the hardest decision is the decision to stop something." When the teams meet together daily, information flows bidirectionally. The meetings are not about the Product Owner or the Business Owner dictating to the Development team. Rather, the development team engages in dialogue, in back-and-forth conversation and discussion and debate on the requirements themselves. The development team does *not* enlist the business owner's help or input in making technical decisions or technology decisions. Instead, what they do is engage in dialogue about the requirements themselves. They seek to understand, they challenge, they push—all in the interest of creating clearer, specific, and, in some cases, different requirements that the business owner recognizes as better features and better functionality.

The development team proposes tweaks and changes to the requirements, coming from a place of understanding how to create and evolve and leverage these requirements for the next iteration of features and functionality. The development team isn't challenging the business owner's customer insight. The business owner is looking at the current challenge from the perspective of what the customer needs today, and what they might need and want tomorrow. The development team is looking at the current challenge from the perspective of how to best solve for the business owner's requirements today in a manner that solves the challenge and that can be extended to solutions that might better solve for what the customer might need and want tomorrow. The development team does more than develop solutions; they partner with the business owner to understand the experience that is important to the customer.

These dialogues require trust. The development teams understand the product not only from the technology and technical perspective, but also, through daily dialogue with the business owner and the product owner, from the perspective of the customer. Through time and habit and practice and commitment to honesty, the joint team creates solutions that are emergent, better than what either team alone would have imagined. The daily meetings are less back-and-forth than they are processual. The meetings are not dictation; the meetings are dialogues.

With this understanding of the product, the development team comes to understand the features and the "why" behind the features. They understand more than "what" it can do; they understand "why" the business owner wants these features. The features are more than simple capabilities. Tied together, they become the experience that the business owner wants to create for the customer.

With this understanding, the development teams learn the product or the feature from the business owner's point of view. They use the features. They understand more than what the feature does; they understand how the feature is intended to be integrated into the product. They understand the role of the feature in the intended customer experience. With this insight, the development team can make recommendations that simplify the experience and enable even more efficient future development. And with the trust that has been established, and with the credibility that the development team has earned by learning the experience of the product, they can challenge the requirements. They can make suggestions and recommendations from the customer perspective, not exclusively from a developer's perspective. They immerse in the product to learn the experience of the product.

Consider:

"If I look back at waterfall projects, we had our business teams working with a requirements team to document requirements. The requirements team would then hand over the documented requirements to a design team. The design team in turn would read the requirements, think through the design and then hand it to developers who would then code it.

"The way it works today—at least some of the time but we're not 100% there yet—is we have the business teams and developers working together on requirements. They're talking through what they want, and our developers hear directly from them, so they're able to flush out those requirements, they're better able to understand really what the business is after, ask questions that they wouldn't necessarily have the opportunity to ask later in the development in stage" (Interview with Jeff Mander).

This is processual leadership—leadership performed by developers and businesspeople together—that generates clearer requirements, and requirements that are prioritized. Jeff contrasted this approach to processing and to generating useful outcomes, to not-so-fond memories of how this worked in the traditional waterfall approach. "I don't know how many times I've heard it in the past where something gets delivered and it didn't meet the customer needs or didn't meet the business partner needs. That pretty much goes away with agile because the teams are so interconnected. We have demos during the process that show them what we're building as we're building it. The design and development go hand in hand because we're building it with them. They're seeing the progress as we're developing. We don't waste time and we don't have delays."

Every leader faces cost challenges. Leaders of agile software development teams are no different. The daily meetings between businesspeople and your software development team can help with cost challenges. Jeff relates his experience with addressing cost challenges in the daily meetings.

Consider:

"As part of our ceremonies within the sprint process, our business partners prioritize the stories themselves. However, there is no value-to-complexity comparison until we hit a constraint: do we want to deliver the code faster, or do we want it to be cheaper? Then we have to start deciding which elements of each story we want to move higher up the priority list and which ones we're willing to move lower down the list."

"For example, one of the projects we worked on recently required a ton of work on the system integration side. We developed the cost estimate, and it was much, much higher than the business team could afford to spend. So, my project owner and my business owner talked, and the team recognized that the requirements were going to cost more than the project could afford, mainly due to three specific requirements. So, we asked if there were requirements could come out. Is there something that could be descoped? If we could make changes to the requirements, then the cost estimate would be cut in half. It's those kinds of conversations that we have every day.'' Understanding the requirements, and as importantly, understanding the *value* of the requirements, helps the developers prioritize each requirement against every other requirement. "We're not there yet, we're not quite where we need to be all the time," Jeff acknowledges. "We want to get to where we're doing story pointing for every new story that comes in. This will help us figure out how complex it appears to be before we start coding it. We want to have our business partners start assigning values to each story so that we can work with them to assess the value versus the complexity. When we reach that point, we'll be doing exactly what we're striving to do: get clear on the value and prioritize our efforts. Having this clarity right at our fingertips helps us see where there might be opportunities to deliver faster or to reduce costs."

Are You Ready?

You started by deconstructing this Principle to get at its meaning. You have defined the words and you understand the words. You've defined your objective with this Principle by starting with the end in mind. You've defined what matters most. You know why this matters most because you've asked great questions of your team, your boss, and your customers. You've asked great questions and listened, really *listened*, to what you were told, and you have learned from their stories.

Now that you are clear on what it means for businesspeople and developers working together daily throughout the project, what do you do? You are the leader. It is time to act.

Do

Connecting to Collaborate

Effective collaboration requires generative relationships, and generative relationships require trust. Gisela Backlander's research "highlights two important implications for managerial practice: (1) fostering opportunity for constructive dialogue, as this is a motor of continuous improvement and change; and (2) the value of human attention to the quality of interactions" (Backlander, 2019).

The most effective means of influencing the structure, processes, and functioning of complex organizations … is to focus on impacting the relationships between agents in the system (Regine & Lewin, 2000; Lewin & Regine, 2001). Relationships are the principal organizing factors in complex organizations where everything exists only in relation to everything else. Weak relationships create negative energy and limit what the organization can accomplish while strong relationships potentially enhance adaptability leading to increased creativity and innovation (Lewis, 1994). Dennis Tourish's research reaches this same conclusion: "From a process perspective, organizations are constituted through the relationships between people" (Tourish, 2019). Positive interactions between agents in a system create a sense of mutuality where trust is increased, and people have greater respect for and impact on one another. Effective interactions affect how people work together and are more likely to lead to collaboration in situations where participants might otherwise be competitors. This sense of mutuality causes people to become increasingly interconnected, and the resulting behaviors are more novel and unpredictable. The rich connections and changing patterns of interactions can lead to the evolution of more robust, adaptable, and successful organizations (Lewin, 1999).

You are a leader doing complex work in a complex organization. Positive relationships are critical. The ability to form and maintain these types of relationships while encouraging others to do so as well helps leaders be more effective. Positive relationships become a source of power, enabling organizations to evolve and adapt because the people in them care more about their work, their co-workers, and their shared purpose. A culture of caring emerges where people are willing to go to great lengths not to let others down, and the resultant behaviors lead to greater creativity, innovation, and productivity (Zimmerman et al., 2008).

Consider:

"For an organization to seek stable equilibrium relationships with an environment which is itself inherently unpredictable is bound to lead to failure. The organizations will build on its strengths, fine-tune its adjustments—and succumb to more innovative rivals. In this environment, successful strategies, especially in the longer-term, do not result from managing an organizational intention and mobilizing around it; instead, they emerge from leading complex and continuing interactions between people" (Schneider et al., 2017; Rosenhead et al., 2019).

When people are connected to others through a shared purpose for the success of an organization, they often become capable of doing more than initially envisioned (Kotter, 2012). They are willing to contribute more to meet the needs of the organization, which leads to greater feelings of personal fulfillment. This sense of community may also result in increased loyalty and a willingness to be more flexible, which in turn helps the organization become more adaptable and resilient. Effective leaders in these situations further this sense of connection by investing their time and effort to build commitment to the vision and ensure that people feel valued through mutual, connected relationships (Lewin, 1999).

You are the leader. What do you do?

You have daily standups.

Who do you include? You include your developers, your product owner, and your business owner.

Why do you have daily standups? What's your objective? What's your intended outcome? It's this: you connect so that you can gain insight, achieve clarity, achieve alignment, establish priorities, and make meaningful progress.

What does working together daily look like? It looks like daily standups. These can be in-person or virtual. Either way, stick to the guiding principles of the daily standup: keep it short, keep it focused, keep it structured, keep it succinct.

Who is in the daily standup?

Your developers. They are in this meeting because they are the people who are entrusted with connecting with the business owner in a meaningful and generative way. They are expected to understand the wants and needs of the business owner.

They are expected to translate what the business owner says to code. They are expected to distill from evolving requirements what the customer most wants and needs. They are expected to transform an ongoing dialogue into a product.

Your product owner. The role of the product owner is a difficult one. Software developers have challenging roles and so do product owners. The best product owners are domain experts, strategic thinkers, and customer advocates. They have decision-making authority. And they feel shared ownership in the output your team is creating. The product owner grooms the request, refines the request, then slots the request into the appropriate sprint for development.

The business owner. The role of the business owner is to make clear to your development team how customers will use the product.

Gaining Insight and Reaching Clarity

Businesspeople and developers work together daily throughout the project to stay connected. Connectedness does not guarantee that you will understand your product owner, but it does improve your chances. Connectedness does not guarantee that you will be able to anticipate what your business owner might want next, or how the product might need to evolve, or how your product owner will respond to the next demo, but it does improve the odds.

Achieving Alignment, Establishing Priorities, and Making Meaningful Progress

When businesspeople are in the room with the development team, are they being listened to, or are they being placated? Is their input being heard, or is it being ignored? Is it acted on, or is it dismissed? When the teams are in the room together, this might look like alignment, by all appearances it might look like the teams are collaborating, but what's really happening? One leader described some of the leadership meetings of department heads as resulting in "pseudo-alignment."

"The leaders might say they're aligned, maybe they really think that they *are* aligned, but are the people defining the requirements and the people coding these requirements *really* aligned? The leaders enter the conference room and discuss and position and exchange, but do they listen? Their teams perceive them as peacekeepers. Do they challenge? Do they disagree? Or do they nod their heads, do they nod a collective nod, and emerge as misaligned and entrenched as ever? Or do they really think they're aligned because they said what they wanted to say, and they heard what they wanted to hear? They think they're aligned; they tell their teams that they're aligned and that there's alignment across the business, but the working teams—the development team, the businesspeople, the product owners—find out differently." One of Amazon's leadership principles applies here: Disagree, then commit. The notion of disagreeing, and then committing, can be a pragmatic approach for causing an honest dialogue because with this notion, you are explicitly stating the expectation that there will be disagreement.

To an extent, the idea of "disagree, and commit" more than makes disagreement safe; it encourages disagreement. The participants are invited to share their views and to be open about what they believe and why they believe it. By establishing this condition—that people in the room will and are expected to disagree—you as the leader are establishing a condition that will foster growth. The views of each person are acknowledged. The person who disagrees isn't bullied into changing his or her mind. He isn't forced to capitulate. She isn't required to agree. The beauty of Amazon's principle is that it allows for disagreement and doesn't require consensus before the team can make progress.

Achieving commitment is the outcome generated by constructive dialogue and a shared objective. Michael J. Arena's and Mary Uhl-Bien's research supports this view. Their research resulted in findings that suggest "what is needed in complex organizations is an adaptive response—one that involves engaging, rather than suppressing, the tension generated in the conflicting perspectives of the operational and entrepreneurial systems" (Arena & Uhl-Bien, 2016).

A notion from complexity theory is at work here. You as the leader encourage flexibility and adaptability rather than exert rigid control. When participants in the dialogue understand that they are permitted to disagree and will not be forced to change their mind or state that they agree when in fact they don't, they are being acknowledged as professionals with legitimate points of view. Their professional credibility isn't at stake if the team ultimately chooses to proceed in a direction that he or she doesn't agree with. This degree of adaptability and flexibility is necessary for progress. If on the other hand the team will only proceed when there is consensus, one of two things will happen: the team will not make progress when they encounter a difficult choice, or they will make progress but at the expense of honest dialogue.

There are various approaches that managers can use to control the amount of tension and ambiguity in an organization. For example, leaders can adjust the extent to which an organization is centralized or decentralized to introduce a greater degree of freedom, which then increases the amount of complexity in a system (Stacey, 1996). The degree of competition and cooperation between individuals or groups in an organization can also be adjusted to raise or lower the amount of tension and anxiety in the system. A higher level of creative tension can be achieved by encouraging people to take risks, not providing them with explicit directions, not taking over control of challenging situations, and by allowing problems to simmer for a while rather than fixing them immediately (Lewin & Regine, 2001) (Kelly & Allison, 1999). Leaders need to learn through experience how to use these levers to achieve the right balance and adjust their managerial approach to the requirements of varied circumstances. This is not easy to do, and it requires constant attention and adjustments. Too much control can lead to rigidity and an inability to respond to challenges except with previously tried approaches. On the other hand, too much stress or freedom can cause an organization to fall over the edge into a chaotic or paralyzed state. Complexity theory encourages the introduction of a healthy level of tension and anxiety into a system to foster flexibility and adaptability that will result in optimal creativity and organizational effectiveness.

Key Takeaways

1. Listen.
2. Collaborate daily with your business owner and your product owner to gain insight and achieve clarity into how the customer will use the product.
3. Meet daily to maintain alignment on priorities. Misalignment can occur. Meet daily to realign.
4. Meet daily with your business owner and your product owner to ensure that your team makes meaningful progress. Because delivering valuable software is the name of the game.

References

Arena, M., & Uhl-Bien, M. (2016). Complexity leadership theory: Shifting from human capital to social capital. *People and Strategy, 39*, 22–27.

Backlander, G. (2019). Doing complexity leadership theory: How agile coaches at Spotify practice enabling leadership. *Create Innovation Management, 28*, 42–60.

Kelly, S., & Allison, M. A. (1999). *The complexity advantage: How the science of complexity can help your business achieve peak performance*. McGraw-Hill.

Kotter, J. (2012). *Leading change*. Harvard University Press.

Lewin, A. Y. (1999). Application of complexity theory to organization science. *Organization Science, 10*(3), 215.

Lewin, R., & Regine, B. (2001). *Weaving complexity and business: Engaging the soul at work*. Cengage Learning.

Lewis, R. (1994). From chaos to complexity: Implications for organizations. *Executive Development, 7*(4), 16–17.

Regine, B., & Lewin, R. (2000). Leading at the edge: How leaders influence complex systems. *Emergence, 2*, 23–52.

Rosenhead, J., Franco, L. A., Grint, K., & Friedland, B. (2019). Complexity theory and leadership practice: A review, a critique, and some recommendations. *The Leadership Quarterly, 30*, 1–25.

Schneider, A., Wickert, C., & Marti, E. (2017). Reducing complexity by creating complexity: A systems theory perspective on how organizations respond to their environments. *Journal of Management Studies, 54*(2), 182–208.

Stacey, R. (1996). Emerging strategies for a chaotic environment. *Long Range Planning, 29*(2), 182–189.

Tourish, D. (2019). Is complexity leadership theory complex enough? A critical appraisal, some modifications and suggestions for further research. *Organization Studies, 40*(2), 219–238.

Zimmerman, B., Lindberg, C., & Plsek, P. (2008). *Edgeware: Lessons from complexity science for health care leaders* (2nd ed.). VHA, Incorporated.

Agile Principle 5: "Build Projects Around Motivated Individuals. Give Them the Environment and Support They Need and Trust Them to Get the Job Done"

Abstract

People matter. Relationships matter. By developing talent, you are creating conditions that foster growth, both personal growth and professional growth. When you empower people, you are creating conditions that foster growth. People who feel empowered feel trusted. People who feel trusted feel empowered. And people who are trusted and empowered will make things happen.

Vignettes from leaders at Centene, Nokia, Amazon, UScellular, and Oracle illustrate the leadership challenges that come with developing talent.

Motivated individuals are key to success. You as the leader motivate and inspire individuals from the day they join your team to the day they leave. You motivate and inspire the people on your team by teaching, sharing, empowering, challenging, trusting, holding accountable, rewarding, and developing.

What Does This Principle Mean?

What does this Principle mean? Here's a way to think about this: People matter most. Listen to them. Care about them. Support them. Trust them. Develop them. The environment you help create matters to everyone.

Let's deconstruct this principle.

"Build projects." What kinds of projects? Projects whose value and meaning are clear. Valuable projects and meaningful projects are worthwhile projects, and worthwhile projects attract and retain people who want to do valuable and meaningful work.

"Motivated individuals." What is a motivated individual? What does he or she do? One: she tries. She makes an effort to seek out information, to seek connections, to seek solutions, and to seek opportunities. Two: she builds and grows her professional network. Three: she commits to continuous learning. Motivated individuals will distinguish themselves. And you as the leader need to recognize them. This is part of identifying talent, which you must do before you can begin to develop talent. So, when you hire, follow Amazon's principle: "Hire and develop the best." Insist on high standards.

"Give them the environment they need." What does "environment" mean? Alex King describes an aspect of an environment that attracts and retains talent. He encourages leaders to ensure that the team has a clear long-term goal for what outcome they want to achieve. Not necessarily the specific product, but a good pulse on what problem they are solving. Having at stable 6-month or 1-year roadmap greatly benefits your team.

"Give them the support they need." What does "support" mean? Support is what enables your team to get today's job done. This is not to be underestimated. The work at hand is the work that needs doing. But the work at hand will end, the code will be committed and delivered, and the customer will operate what the team has created. What's next? Without a roadmap and without a commitment to developing your talent, you as the leader will find yourself delivering iterative support, the same support each time around, the same support for tomorrow's project as you provided for yesterday's project. You as the leader must support your team in getting today's project delivered, *and* you must develop your team, so they are ready for tomorrow's project.

"Trust them." What does "trust" look like? Extending and demonstrating trust. You share. You empower. You permit. You allow. You accept. You develop. You trust. You encourage.

You Are the Leader. What Do You Do?

Start with the End in Mind

What will this look like when you've achieved it? When this Principle is in place, your team will be inspired and engaged. They will do great work. They will know that someone believes in them. They will know they are empowered. They will know they are trusted. They will thrive. They will know they have the opportunity and the support to do great work in service of something bigger than themselves.

Listen and Learn from Others

How do you get to the "end in mind"? You start by asking great questions, and then you listen.

- What motivates an individual?
- What motivates you?
- What environment do you need?
- What does "support" look like to you?

- What does "trust" look like to you?
- How can I help?

We've started by deconstructing this Principle to get at its meaning. We have defined the words and we understand the words. We've defined the objective with this Principle by starting with the end in mind. We've asked great questions and listened, really *listened*, to what was said. Now it's time to learn from others.

"Build Projects"

As experienced leaders, we all know that projects come in different shapes and sizes and complexity and duration. What I'll define as the "best" projects are those with social value. What kinds of project have social value? Mapping genomes, for sure. Desalinization to create drinkable water and depolymerization to rid our water and our land of plastic. In terms of software development projects with social value: search engines, health applications, platforms that connect people in ways that matter to them.

Meaningful work has social value. The software development that enables meaningful work is by extension meaningful work in and of itself. Developing the platforms for measuring fitness; for enabling connections between people; for designing safer roadways; for making flying safer. All of these have social value. And technical people want to do work that matters. We are fortunate as leaders if we get to participate in work that has social value. Teams will feel more motivated to work on those projects that they see meaning in. You as the leader must be able to define and describe the value in the work.

Make it clear that the work matters. If you can't make it clear, then does the work really matter? And if the work doesn't really matter, why are you asking your team to do it?

"Motivated Individuals"

What is a "motivated individual"?

What is a motivated individual? How do you recognize a motivated individual? A person who is motivated demonstrates an enthusiasm for doing the work. I want motivated individuals on my team. I want people who show up every day with the internal drive to do great work. If they're not motivated, if they don't have the internal drive to do great work, I don't want them on my team. You might be thinking, "Well, it's the leader's job to motivate her employees." I disagree. The leader's job is not to cause a person to feel ownership and pride and energy and drive; that is up to the individual. The leader's job is to provide meaningful work in a generative environment. It's to provide meaningful work that inspires motivated people. Make the environment generative. Below I'll describe how you as the leader can do this.

If I have to motivate you, I'm spending too much leadership time and leadership energy on the raw materials. Each employee needs to bring the raw materials: the energy, the enthusiasm, the professionalism, the drive. The leader creates an environment where people can do great work in service of something bigger than themselves. This environment includes projects that are meaningful and valuable. The leader brings the work to the right people. When I think about motivating people, I think about cheerleaders at a high school football game. They stand on the sideline and dance and yell and exhort the crowd to be enthusiastic, to wave their arms, to stand, to clap. People have filled the stands, they've turned up, they've taken their seats. And the cheerleaders try to get them engaged in the game. This is what you do for people who are on the sidelines and in the stands. This is what you do for people who are not *in* the game.

The leader of a team should not have to do this. The leader should not have to dance and cheer and wave his arms to get his team to want to do great work. Every person on every team is personally responsible not simply for showing up but for entering the day and approaching the work as paid professionals committed to doing great work. Give the leader a motivated team and then require that the leader provide the inspiration. There are many ways a leader can inspire a team. Show your own commitment. Show your own excitement. Share why you care. Connect their work to an even greater cause. Explain why customers or stakeholders care about their work. Demonstrate how their work matters to their customers.

Let's consider changing the word "motivated" to "inspired." Here's why: if your challenge as the leader is to motivate people, I would argue that you've got the wrong people. People who need daily motivating require a cheerleader; your role as the leader is not to lead cheers. Your role is to create an environment where each individual can do great work in service of something bigger than themselves. Creating such an environment requires that you as the leader have and can articulate a compelling vision of the future and a compelling vision of the value of the work that you're asking your team to do. Creating such an environment requires that you as the leader can articulate the value that each person on the team brings to that team. Creating such an environment requires that you as the leader understand what "value" means to each person on your team, how they define it, how they view it, where they see it, and where they don't see it. You as the leader need to know, and *enable*, each person on your team to connect themselves to the value of the work in front of them. That's inspirational leadership. The people that you want on your team are people who find the motivation inside themselves, and who are inspired by an environment where they are enabled and empowered to do great work. The definition of great work varies from person to person, but includes these components: it's meaningful, it's valuable, it's durable, it's high-quality, it makes a positive difference in the work or the life of someone else.

How do you inspire individuals?

We all want inspired employees on our teams. We want to work with motivated employees on other teams. How do we find them? Where do we get them? How do we create them, and how do we sustain them? If you're lucky, you've inherited a team of inspired individuals. We should all be so lucky! More likely, you as the leader will have to do the hard work, the creative work, the ongoing work, of creating an environment that inspires people.

In this next section, I'll show how to create an environment that your best employees will find inspiring. In the next section, I'll describe how to give these inspired employees the support they need.

So how do you create an environment that your employees will find inspiring?

Instill an owner's mindset.

Most employees work because they need to work. This doesn't mean that satisfaction doesn't matter. It does, of course. Employees want to find satisfaction in their work, much as you do. One way that people find satisfaction in their work is to see and to experience that they have a stake in the success of the outcomes. What does this "stake" look like? Bonuses. Stock awards. Equity. Profit sharing.

Doug Lowell said it well. "Get them to behave, not like an employee, but like an owner. Give them a big enough stake in the outcome that they feel like they own their piece of this and not that they're a cog in a machine." Without the reward of the stake in the owner's outcome, then it's lip service. What you do is up to you. But choose *something*. Instill an owner's mindset.

Provide meaningful work.

You as the leader of your team must be able to articulate the higher purpose for the team's work. Provide an environment where the work is meaningful—intentionally and explicitly connected to a higher cause—and where the individual's connection to meaningful work is made explicit.

Two truths about technical teams: (1) They want to do work that matters, and (2) they want to work with other smart people. To attract talent to your team, to retain that talent, and to inspire that talent, you must articulate a higher purpose for the work you are asking them to do. Everyone on your team needs to see that they are part of something bigger than themselves. And one rule about smart people: Smart people want to work with smart people.

To optimize the performance of an organization in a complex organization doing complex work, leaders must take responsibility for providing their people with fulfilling work that enables them to reach their full potential. They should create conditions that enable people to develop and grow by fully engaging them in the workplace. By assigning responsibilities that accommodate each person's interests and skills rather than merely assigning them to a job, leaders can release the energy in each person that is the difference between doing just what they need to do to stay employed and doing whatever is necessary to meet a challenge (Levy, 2000).

How does this tie to Principle 5? You as the leader must "build projects"—bring meaningful work—to your team. If you've been able to articulate the value of the work, and the role each person plays in bringing that meaningful work to life, you've done good work. The people you want on your team are those people who are internally motivated by the prospect of doing meaningful work, of being trusted and empowered and supported to do this meaningful work. People who are not internally motivated and who are not inspired by the opportunity you create and the environment you create, they may not be the people you want on your team. People who need daily motivating, daily pep talks, will drain your energy and will distract you from the work of the leader. People who aspire to do great work, people who are inspired by the opportunity to contribute, to make a difference, to work with like-minded colleagues, these are the people who are more likely to bring their best selves to their work. Treat each team member as a professional and reward them as professionals. Those who are internally motivated and who are inspired by the opportunity and by the environment and by the support you provide will bring their best selves to their work.

How do you inspire people: give them meaningful work, show you trust them, empower them, and create opportunities for them to do their best work, and create opportunities for them to realize and achieve their potential.

Motivated individuals should be the ante, the ticket to entry, the admission criteria to join your team. But motivation is not enough. It's necessary to be sure, but it is not enough to differentiate your team and your company from your competitors. Start with people who are driven. Start with people who are internally motivated. Accept nothing less than people who are motivated and driven. There's no shortage of motivated people. You'll find them across industries, and you'll find them up and down and across the organization. Motivation isn't the problem, and it isn't the answer. The answer is *talent*.

"Give Them the Environment They Need"

This includes the physical environment and the professional environment.

Physical environment.

Doug Lowell shared how he and his team recognize and respect the importance of the physical environment. "You have to get away from the 'penny wise, pound foolish' approach that prescribes that everybody gets one laptop and two monitors and one cell phone. If people say they could benefit and work faster with something else, give it to them. Of course, there are realistic limits like budgets, but I think we get way too focused on the small decisions and the constraints, and we limit people's creativity and productivity."

Doug shared his experiences with the design of a new office building.

Consider:

"Not too far in, we realized we didn't really know who we were building it for or how it was going to be used. So, we switched gears to inject flexibility into as many aspects of it as we could. We shifted our approach and instead designed for software developers and technologists. The IT leadership asked for an environment that would become an extension of their creative process, so our philosophy and our approach changed. We created a flexible and modular physical environment where the teams would feel free to move things around and reconfigure the space. We made the workspace work for their process. If it wasn't working, they could rearrange it and do something different. The physical environment became a way of management messaging to the team: 'We really do want you to experiment. We want you to break rules. We want you to think out-of-the-box and color outside the lines.' The way the work environment was delivered was a very direct way of messaging that because it was obviously meant to be 'torn down' and rebuilt, meant to be reconfigured and moved, like Legos. That's how we used the physical environment to reinforce the message that we wanted them to be innovators and people who thought differently and would make whatever changes they needed to foster their process."

Professional environment.

Set meaningful and achievable goals.

You as the leader need to ensure that your team's goals are aligned with the goals of the teams you support. Be clear what the goal is, and why the goal is important. If your marketing team wants to drive traffic through your ecommerce channel, and the Sales team wants to drive online sales, your development team will be charged with developing one-click purchasing on your company's website or mobile application. You need to ensure that your team designs and delivers in a manner that supports the Marketing and the Sales team's goals.

What about stretch goals? What about stretch objectives? If you believe that your role as the leader is to set an unrealistic goal, your team will feel defeated from the outset. But what if you could get your team approaching problem-solving differently? Pose the challenge, share the compelling 'why,' and see what they can do. Help get your team to the point where, when they're confronted with a particularly daunting challenge, they say, 'That sounds impossible. It sounds like it can't be done. But if it could be done, what would it look like? How could we solve this, even partially?' In Doug's experience, "that's a way of getting a team to aim high and to do so in a way that they break down the problem into small enough pieces that they really do start solving it in creative ways."

Establish clear expectations.

One of the most important aspects when you set expectations is that you hold your team accountable to those expectations. Be clear with your team what you expect of them and how you will be evaluating them. Do this not only for project work but also be clear with them how you evaluate their performance.

Practice clear and consistent accountability.

Be relentless but not ruthless.

One way to think of what accountability means: it is meeting your commitments in the eyes of others. Also: Think of accountability not as punitive but as empowering. This includes being clear on what each individual, and the team as a *team*, need to demonstrate and what they need to deliver.

Share information.

Conduct open forums. These are small group meetings—8–12 people—that are designed for sharing information. You share information with your team. You listen. You don't assign work; you don't ask for project updates. You are sharing information that the people in the room would not get otherwise. You share insights from senior leader discussions. You share perspectives on industry trends. You share a view on how you see the marketplace evolving.

In these forums, not only are you sharing information and listening, but you are also teaching. This is an opportunity for you to share stories. Share the company's story. Share *your* story. You are the leader, and you have stories about your career, about your development, about leaders you've had. You have stories about what you've learned and how you've learned. Stories are one of the most powerful ways to connect that human beings connect. Share your story.

This approach is supported in the academic research. Boal and Schultz "put forward a view that places dialogue and storytelling as a central mechanism explaining how strategic leaders are involved in the emergence of new behaviors in complex adaptive systems" (Rosenhead et al., 2019).

Foster relationships.

Your direct reports must build relationships.

The environment that your team needs, and the support that they need, does not come only from you. You alone cannot provide the fulsome developmental and supportive environment that will enable the employees on your team to do their best work and to reach their full potential. Each person on your team needs to be connected to other people, not only on their own team, but throughout their business unit and throughout the organization. Help them recognize the value of a professional network and help them build those relationships.

Employees who feel supported will be more motivated. They will recognize that they can count on others and, perhaps even more importantly, that others count on them. As with mentoring relationships, the most generative, valuable, and nurturing relationships in a professional network are bidirectional.

Relationships matter throughout your employees' tenure with your team and throughout their career. You as the leader must start early, educating your team on the value of relationships, and demonstrating to them the value of your own professional network.

I've witnessed the value of networks for employees of all tenures, from a team of summer interns to a newly appointed CEO. One particularly striking example comes from a team of summer interns. This team shared with me their experience with the value of building relationships and the value of networking. The value was clear, and the lessons were valuable. Each intern worked in a different team and had a different leader for their summer assignment. But what they had in common was a nurturing leader who right from the start encouraged them to develop networks.

In fact, establishing connections, establishing a network first in their team and then across the IT organization, was a specific deliverable that each leader wrote down and included in each intern's work plan. In addition to assignments to learn different tools and to learn how to apply those tools in developing software solutions, each intern was assigned to identify colleagues who would become a part of their internal network. They each did this, and they reported near the end of their assignment that the value of their network exceeded their expectations. They became connected with colleagues who could teach them technical skills, who could teach them not only *what* to do but *how* to do it, how to approach problem-solving, how to navigate the organization, how to "connect the dots," and see the connection of their work to the higher-order objectives of the company. Each intern was able to connect his or her summer assignment to specific value for the company. One intern created an automation for a discrete activity for customer service agents; this activity freed up the agent to spend time on more complicated calls with customers who had more complicated issues. Another intern automated the creation of test scripts for bimonthly software releases. He focused on automating a task that could take several testers hundreds of hours each cycle, and he created an automated process that now takes under 10 h. The environment that the leaders created, and the support that they provided the summer interns, helped create a memorable and positive experience for these interns. They were assigned real work with real value, and they completed their internships feeling they had made a difference.

Relationship-building and network-building serves interns by introducing them to an organization, its teams, its structure, its interconnectedness. Relationship-building and network-building for tenure employees serves a different purpose. Tenured associates develop networks and form relationships to learn, to share, to expand their sphere of influence, to grow as enterprise thinkers by learning what concerns the employees and leaders in other parts of the company, and how those employees and leaders make decisions. In building relationships and building networks, the number of people—the nodes on the network, if you will—matters, but what matters even more is the quality of those connections.

Encourage and assist your employees in establishing meaningful connections with employees and leaders in their immediate circle and with more distant stakeholders. The nature of the relationships should create the opportunity for your employee to build and establish herself as a trusted colleague, a valuable resource, and a direct, transparent, productive, and honest member of the team.

Instruct your employees to pay close attention to the style of the interaction with each of the leaders they connect with: if they connect with the CFO, your employee should pay close attention to the kinds of questions the CFO asks, the concerns she expresses, the objectives she's most interested in. Your employee will benefit from learning how senior leaders think, and this is only done by building and nurturing a relationship, and by being purposeful in each interaction. Your employee will heighten their effectiveness in working "up" the organization by learning and adapting to the senior leaders' styles of communication, not only listening to what they say but listening for how they think.

Here's what you do as their leader: connect each person on your team with a person from Marketing, a person from Finance, a person from Sales, a person from Customer Service, and a person from Corporate Communications or Public Relations. The point is not to overwhelm anyone on your team, and the point most certainly is not to distract them from their primary role. These are connections that should be nurtured over time. And make it clear to your team that each relationship needs to be bidirectional. There should be give-and-take, teach-and-learn, so that your team and their connections in the other parts of the company learn from each other and strengthen each other.

Create a safe space for your team. You can talk about psychological safety all day. But there's nothing more powerful, nothing more convincing, than demonstrating that you've created a safe space. One way to do this: admit your own mistakes and share the impacts of your mistakes and the outcomes from your mistakes. Did the product fail? Did you learn from what happened? Share honestly and openly. Your examples speak loudly. They reveal not only your humility, but they also set the example for your team. They too will make mistakes, like you have. They too will be concerned or worried or scared, like you were. Set the example that mistakes happen—you make them, they make them—and that as a team, you discuss them, learn from them, and you move on together.

Set the example. Be the proof that people make mistakes and admit their mistakes and survive their mistakes. You as the leader will make them. And your team will make them. Work, and life, go on. Clint Wallin shared his advice on how to treat mistakes. "You're going to make mistakes. And you have to be able to get over that. You have to be able to get over it. You can't let the mistake that you made yesterday hold you back from leading your team well today. You've got to move forward. You must be able to forgive yourself. Learn from it and move on. There's no time to dwell. It's too fast-paced."

"Give Them the Support They Need"
"Support" is a big word. Include in it your efforts to help your employees improve their personal effectiveness. Personal effectiveness includes knowing how to challenge an idea, and how to show up as a valued and valuable teammate.

What does "support" look like? It is more than sharing information and clearing hurdles. It is more than helping get questions answered. It is more than procuring tools. Top-notch support is far more. It includes each of these, but the best leaders and the most effective leaders know there is more to support than these. Sharing information with your team is a form of support. Creating development plans, providing feedback, helping chart a career growth plan, all are forms of support.

A key differentiator between good leaders and the best leaders is this: good leaders *support* their teams, where the best leaders *develop* their teams. Good leaders procure tools for their teams; the best leaders provide resources. A good leader tells their teams why they need to do what she's asking them to do; the best leader provides a compelling "why" for the work she's asking her team to do. Good leaders tell their teams that their work is important; the best leaders demonstrate the value of their work. Lastly, and arguably most importantly, good leaders help each person on her team to get their work done; the best leaders invest themselves in developing each individual.

Supporting the whole person.

At work

Support your employees in taking the time necessary to build their skills and capabilities. Don't rush their development. Understand each of your team members well enough so that you can push them in a way that challenges them to grow, but that doesn't overwhelm them. I had a boss who told me early in my first executive role that "if you don't stretch, you don't grow." That was true for me, and I believe it's true for most people. Your responsibility as the leader is to work with each of your employees to find that balance between healthy challenge and demoralizing challenge. You can find that balance only if you invest the time and attention to understand how they do their best work.

Seeking to understand is not the same as prying. We sometimes hear that as leaders, we need to learn everything there is to learn and know everything there is to know about our employees. Not true. We don't. We don't need to know any more about an employee's personal life than they are willing to share. Demonstrating that you care can be done by demonstrating empathy. And empathy doesn't require that we pry, only that we listen attentively.

When you listen, you'll hear that one of your employees has elderly parents and that she spends time on the weekends caring for them. Or you'll hear that he is integrated into his community through service at the local Boys and Girls Club and so is less likely to be willing to relocate and give up those connections. Or you'll hear that she has a passion for solving puzzles or riding motorcycles or gardening. You'll be able to connect in genuine ways when you invest in listening, really *listening*.

For those people on your team who aspire to more senior roles, make it clear that demonstrating patience is acceptable, even desirable. Most of us are familiar with the Peter Principle. The principle states that in any organization, employees tend to rise to his or her level of incompetence. This principle was originally intended as a satirical jab at leaders in organizations, but there is wisdom in it. No employee wants to feel that she is incompetent in her role. To feel capable and to perform capably, people need time to grow and to develop. Push people to help them move them, don't push them and cause them to move out.

Make it clear to your team that patience can be valuable. Patience isn't the same as laziness or lack of direction or lack of drive. Rather, patience can serve an ambitious employee. It takes patience to understand and accept the developmental assignments, to work through them, to demonstrate success and learning. It takes time to work a plan. It takes patience and it takes perseverance. The objective of development is not to be done with it; the objective is to *develop*.

Work with your employees. Take the time to create a development plan that allows for valuable growth. People grow when they are challenged with diverse experiences and developmental opportunities on real and important projects. Set aggressive targets for completing the work and demonstrating the development; again, not so aggressive that the employee is likely to fail and miss out on the growth opportunity but aggressive enough that the employee is challenged to apply themselves and stretch themselves. John Wooden, the legendary UCLA basketball coach, said it well: "Be quick but don't hurry." Work the plan diligently and develop purposefully.

Encourage each person on your team to find a community. Why? Because communities can be sources of security, of challenge, of confidence, of identity, and of support. Communities listen. Communities care. By definition, a community is "a feeling of fellowship with others, as a result of sharing common attitudes, interests, and goals" (Oxford Languages). A person's community is one more source of support. Whether the community is an affinity group, a developer team, or alumni from a local college or university, it can serve as an anchor, a "home base," and a source for encouragement.

At home

Recognize that like you, each of your employees has a personal life. Many have family lives. Be careful not to underestimate the significance of a person's personal life on their work performance. A myriad of issues can impact employees: social issues, like civil injustice or civil unrest; political issues, like decisions or choices or decrees; environmental issues, from severe storms to climate change; and of course, personal issues such as the challenges that come with young children or aging parents, or a spouse who needs to relocate for work or for family reasons, and mental or physical health challenges. You need to know your team well enough to know when to give space, when to ask, when to listen, and when to provide time and space for them to express themselves and care for themselves. They may need space to reflect, they may want to talk, they may need to feel heard. Understand your team well enough to understand how best to support them.

You as the leader need to recognize that like you, the ways in which each person on your team feels these issues and is impacted by these issues will vary. Your role as the leader is not to solve these issues; rather, your role is to *recognize* that each person will respond in his or her own way. Your role is not to interfere with their response or try to change their response; your role is to *respect* that each person will respond in his or her own way. Your role as the leader is not to judge; your role is to *appreciate* the stress or the strain or the anxiety or the distraction that life issues can cause. You need to know your team well enough to know how best to respond.

Giving your team the support that they need extends beyond support in how they do their job and how they grow develop and grow professionally. Giving them support they need extends to recognizing each person as an individual with complex thoughts, feelings, and emotions.

Remember, each person, like you, contains multitudes.

"Trust Them to Get the Job Done"

One of the ways you demonstrate trust is by sharing information with your team. Keep people informed. Keep people connected. Keep people aware. Share the context, the status, the performance, the forecast, the trends, the wins, and the losses. Share.

Now, if someone on your team demonstrates that they cannot be trusted, get rid of them. Seek to understand what happened, and once you do, don't hesitate on your next step. And your next step depends on what you learned when you sought to understand. If the issue was an honest mistake or misinterpretation or miscommunication, recognize that, acknowledge that, and correct for that. If the incident was one of bad behavior, egregiously poor judgment, lying, stealing—get rid of that person. Either way, whether you learn that it was an acceptable issue or an unacceptable issue, don't hesitate to act.

Mike Irizarry shared his view on how to build trust.

Consider:

"As the leader, you've got to make deposits into that trust account through your behavior and your conduct. And I find that one of the greatest ways to start building trust is to share your vulnerabilities. When you admit that you've made a mistake. When you acknowledge that there's something you don't know how to do. When you ask for help. Doing that builds that deep, deep trust, not the superficial trust that we hear about, but the deep trust that forges a team."

Consider the oft-told story about the drivers at a package delivery company being told how to carry their keys so that they could more quickly, get back in the van, and get it started and get it rolling.

Consider:

"Those directives were effective to a degree. But imagine if you let the drivers figure out for themselves. What is the most effective, efficient way of getting the packages delivered? It's the difference between hiring people who will be robots and in which case you should just make a robot to do processes that you've already figured out and you're convinced that you've optimized versus maybe this isn't optimal and maybe somebody can figure out a better way to do this" (Interview with Doug Lowell, 7/21/22).

Are You Ready?

You started by deconstructing this Principle to get at its meaning. You have defined the words, and you understand the words. You've defined your objective with this Principle by starting with the end in mind. You've defined what matters most. You know why this matters most because you've asked great questions of your team, your boss, and your customers. You've asked great questions and listened, really *listened*, to what you were told, and you have learned from their stories.

Now that you are clear on what it means to build projects around motivated individuals, and what it means to give them the environment and support they need, and what it means to trust them to get the job done, what do you do? You are the leader. It is time to act.

Do

Ideas are not today's differentiator. The "next big thing" isn't the difference-maker. Ideas and projects and environments all matter, to be sure. But ideas and environments and inventions and products and features can be copied and duplicated rapidly in today's environment. Ideas, as Michael Dell said to me, are today's commodity. I couldn't agree more. As lots of people are motivated, lots of people have lots of ideas. Ideas don't differentiate good companies from great companies; or if they do, the difference and the advantage doesn't last long. Lasting difference, lasting value, is created by people. It is created by talent. Talent makes the difference. You are the leader. The talent on your team and the talent in your company is your responsibility.

Talent

Attracting Talent

What is your company's value proposition? What is your company's mission? What are your company's values? What is your company's reason for being? And what is your team's reason for being? What do you as the leader stand for? What would the people around you say you stand for? I ask these questions, and you need to ask these questions, because these are the questions that talented individuals want answered.

Today's worker wants to know. Today's work cares about meaning and value and social worth. Workers jump from company to company for more than just pay. Today's employee wants and expects and will thrive in an environment where the work is meaningful—intentionally and explicitly connected to a higher cause—and where the individual's connection to meaningful work is made explicit.

You are the leader, and hiring decisions are among the most important decisions you will make. When you get the people part right, everything else gets easier. And remember, in the words of Cece Stewart, that "you're only as good as the people you surround yourself with, so be very, very intentional about your direct leadership team and who they are. Do they have the right mix of backgrounds and experiences, and will they challenge you? Your direct report team, your leadership team, is so incredibly important. You're only as good as the people you surround yourself with. Period."

Hiring Talent

You want to work for a good person. Your employees want to work for a good person. You want a good boss. Your employees want a good boss. Life is too short, and careers are too long, to work for someone you don't like. You know this. Your employees know this. Be a good person and hire good people. Hire people who can do, will do it, and who fit your culture.

Mike Irizarry discussed what he looks for when he's looking for technical talent. Consider:

"When we bring people into the technology team, we certainly interview for a lot of things. There are soft things like, 'Can they work well with others? Can they collaborate? Can they follow?' Technology is never stationary. It changes. The rate of change of technology is accelerating. So, I expect people to be experts and then to stay experts. And that means that they're always refreshing their understanding of technology, not just for the sake of the technology, but so that they think about and understand ways to apply technologies to solve business problems."

And Doug Lowell's view on hiring talent:
Consider:

"Hiring people who are different than you? That's easier said than done. A lot of leaders believe that effective performers look and behave and think like they do. It's much easier to stay in a comfort zone as a manager and hire people who you think are smart because they think like you do, rather than seeing that they are smart because they come up with innovative ideas that you would not have come up with yourself."

Identifying Talent

Identifying technical talent.

Who has the technical capabilities to develop into the technical experts you need on your team? The capabilities and the skills necessary to develop into a strong technical performer can be identified in several ways. Here's the approach that Hamdy Farid applies.

First, you must have a pipeline, or a funnel, of talent that you can begin assessing. The funnel or pipeline—the talent pool—needs to be reasonably large, because, as Hamdy relates from his own experiences with talent pools, you likely will find not more than 5–10% of the people who perform *very* well.

There are many different sources of talent: internal technical teams; internal talent that can be sourced from nontechnical internal teams (that person who is a diamond in the rough, who is chomping at the bit to move into the technology part of your business); talent from other companies; college interns; college graduates; military veterans. You can source these people through your own network, the networks of trusted colleagues, internal talent sourcing teams from your Human Resources organization, professional technical recruiters, industry conferences, and job fairs.

Next, immerse candidates from the funnel into real-world experiences. Real experiences, real technical problems are far better experiences for assessing a person's talent and identifying who does in fact have the technical talent you need. Real problems, real challenges, and real development projects provide the real scenarios that technical people operate in. Staged exercises and fabricated scenarios and mock drills and simulations have limited value. They do test a person's technical skills. But they are far removed from testing a person's skills in the real world of ambiguity, competing demands, pressure to deliver, and resource constraints.

One type of real-world experience that Hamdy has used is immersing talented technical people in delivery teams. Why?

Consider:

"I'm putting them in the middle of the problem to see how they handle themselves." Not a staged exercise, but a real problem with real constraints. "People who go to engineering school like to solve problems. That's why they went into engineering school and that's why they survived engineering school. They like to solve problems.

"So, if they survive engineering school, they've demonstrated that the capabilities of learning are there. But how will they manage themselves under stress? How will he or she solve the problem quickly? How has he or she equipped themselves with solving the problem? How many of them were actually there in the middle of it, or how many actually solved it?"

Identifying business talent

Hamdy continues.

Consider:

"On the business side, same situation. Throwing them a hard business problem and see how they handle it. Of course, not all the business problems are solvable compared to engineering problems.

"In one sense, engineering problems can be easy. You're dealing with physics; you are dealing with code. You can solve just about any problem, but business or financial problems sometimes are trickier. You have to find the closest answer. So, to identify a business-related talent, I focus more on their thinking process. I'll have the person explain a solution to me. I'm more interested in, 'How did you get to that answer? What was the thinking process that led you to that solution?'"

Consider this dialogue with Heather Ackenhusen, a former leader at Amazon.

Consider:

"Developers and engineers could switch teams whenever they wanted and as often as they wanted. This puts additional pressure on the leader, who is accountable for delivering a product or a service or a feature, to create an environment that would attract and retain top talent. The freedom that developers had forced leaders to get clear on the environment they were creating. That freedom forced leaders to understand and articulate the challenges and the opportunities they were creating. Am I offering interesting challenges, and am I giving the team the autonomy they want and need to be successful in doing their job? Is what I'm asking them to do going to serve the company, and will it also serve the careers interests of developers and engineers? Will they see and believe in the value of the work they get to do; will they be contributing to something valuable? And will they continue to learn and develop the skills that they want to learn and develop as they move through their career? Do they get opportunities to work with more senior technical people who they can learn from?

"As the leader, you have to know how each person on your team wants to grow. You have to understand their career growth expectations. What are the competencies they want to learn? What are their expectations for development and for getting to the next level in their career journey?

"As the leader, I needed to deliver my product, and at the same time retain the talent that I needed to get the work done. One of the most challenging things I found as a leader was having this tremendous, highly talented organization, and doing all I could to retain them."

How did Heather do this at Amazon? In Heather's view, there are three things the leader must do: build a team, demonstrate the higher purpose of their work, and trust them.

Consider:

"You need to build a team of individuals who respect each other. Smart people like to work with smart people. Technical people usually want to work with strong technical people. There's got to be mutual respect, and they've got to demonstrate that they're good collaborators, that they have strong working relationships. To help with that, we'd do a lot of what I would call 'fun activities' to build the sense of team."

"Second: I believe that people want to be part of something bigger than themselves. They want to ask, and they want the answer: 'What's the higher purpose of the work I'm doing? Am I connected to that higher purpose? *How* am I connected to that higher purpose?'"

"The stories are legendary: a customer would send a note to Jeff Bezos telling him how Alexa saved their life. They'd fallen or hurt themselves or needed help, and they used Alexa to call their wife or they husband or their kid, and they were saved. These stories inspire people. They can connect the work they're doing to someone's *life*."

"The third thing is trust. Show that you trust your team. People want to feel trusted. They want their colleagues to trust them, and they want to know that you, their leader, trust them. People want to feel trusted to work hard and to do the right things."

"The people on your team also want you as the leader to know that they care about the customer as much as you do, that they are as obsessed with the customer experience as you are. Whether they're software developers or data engineers or quality assurance testers, demonstrate that you know that they are as obsessed with the customer experience as you are."

You as the leader need to demonstrate that you trust that your team cares as much as you do. I've seen leaders make the mistake of thinking that no one cares about the customers as much as they do. Here's what that looked like: the leader would get frustrated and sometimes even angry because she believed that people were not showing what, to her, was the right sense of urgency. People weren't moving fast enough; they weren't outwardly demonstrating to the leader's satisfaction that the problem or the defect or the outage was important. So, the leader got frustrated and said things like, "You guys, this is *important*. Customers are having a *problem*."

Trust me: people who care don't need to hear this. They don't need to have their commitment to the customer second-guessed. Simply because a person needs time to think through the solution, doesn't mean they don't understand the importance of the problem. And along the way, people will make mistakes. Again, this doesn't mean that the developer or the engineer or the operations technician doesn't care or doesn't understand or appreciate the importance of the work or the importance of solving the customer's problem. Heather: "People are going to make mistakes. But they're going to learn from those mistakes and grow as a person and grow their skills and grow in a direction that they want to move." Whether that's further growing their technical acumen, maybe becoming a lead for the group, running design reviews, whatever it might be, they want the space and they *need* the space and the support to grow. No one grows without making mistakes. You don't stretch, you don't grow.

So exactly what is "development"? It includes each of these:

Equipping Talent with Tools, with Information, and with Resources

Tools
Support your team with tools.
Consider:
"From an IT standpoint, the tools are changing. We have more tools than we ever had before. Each has a specific purpose, and they do great things for what we need to do for development, whether it be scanning our code for security vulnerabilities that otherwise wouldn't be caught until way later in the process or having auto-build processes that will tell you where something might break when you try to deploy it to production. There are so many different tools, but that becomes challenging because now our teams need to learn all these tools and have some sort of support

for those tools, whether it be administration or setup or configuration or even coding within them." (Interview with Jeff Mander.)

Your role as the leader, then, is to listen to what your team tells you they need, understand why, then use your access to resources to procure the necessarily tools for your team. You as the leader also apply judgment here so that you procure wisely.

Information

You as the leader have access to information that your associates may not have. Jeff Mander understands the value of this information and how to apply this information in ways that help his web development team. "With the data we have at our fingertips, we can make much better strategic decisions on what we work on and when we work on it. We're in a position to sort out priorities from noise." Sharing that information and sharing that insight with the team better equips the team to focus on what matters more at any given point in the development process.

Resources

You as the leader do more than get your team the tools they need. You also make available to your team the resources they can leverage to expand their skills and deepen their knowledge. Jeff leverages the changing technology to do this for his team. "There are so many ways for our associates to learn now. We don't need to send them to expensive week-long classes to learn technologies. We can learn via YouTube. We can learn via Google. We can learn via classes like Udemy that are hour-long and virtual. From a technology standpoint, technology has enabled this democratization of information for learning."

Developing Talent

Before you can start developing the talent on your team, you as the leader need to develop the skill for identifying talent. Before you invest your time and the time and resources of others throughout the organization, how do you determine *who* you will invest in? How can you be sure that you are focusing your talent development efforts on the right people? How can you determine, before you begin the hard work of developing talent, who in fact *has* talent?

The short answer is you can't be sure. The better answer is, you've got to establish structured approaches for identifying who does and who does not have the capabilities to develop and grow into more skillful contributors to your team and to the company. You as the leader must recognize and respect and act on "actors' situational knowledge, perceptiveness, and creative potential...Rather than seeing agents in a 'trivialized' manner (namely, as mere carriers of structural forces), conjunctive thinking views agents as being capable of undertaking novel ('nontrivial') action" (Tsoukas, 2017). Let's go through the steps you as the leader should take in developing talent.

Empowering

In most organizations, control is a key source of power that most managers are hesitant to give up. To effectively manage complex organizations, leaders need to develop confidence in their people and believe that they will come up with solutions that are at least as good as what the leaders could do on their own (Stacey, 1996). Relinquishing this control is one of the most difficult things for managers to do. The challenge is to not revert to the old command-and-control approach, especially when the situation seems chaotic and not likely to produce a workable solution (Lewin, 1999). The key is to create a strong ethical foundation and vision to provide support and direction in times of uncertainty and stress. Leaders should make sure people have a clear understanding of the issues and then allow them to address the challenges with minimal direction or guidance.

If you're going to empower your team, then you need to trust them. Empowerment without trust is not genuine empowerment. In Heather's experience, "At the end of the day, they're probably going to make the right decision." Trust them to do that. Trust them to make the right decision. Tell them you trust them. When they do make the right decision, call that out. Recognize it. Whether it's a small win or a big win is less important than the fact that it's a win. Recognize it for what it is. Create examples of wins and what wins look like. Build examples of good or right decisions, and what those good and right decisions look like. Understand and share how the team got to the right decision. Then, when a decision turns out to be wrong or turns out not to be a good decision, call it what it is. And as you did with the right decisions, understand how the person or the team got to that decision, and what made it a wrong or bad decision. You're not judging people. You're judging decisions. Make it clear that your critique is not personal. Celebrate, and you'll create confidence. Correct and redirect, you'll create trust. Criticize or mock or belittle or blame, and you lose your team.

Making Mistakes

Mistakes. People will make mistakes and they will learn from those mistakes. What lessons will they find in their mistakes? You as the leader help determine that. Will they learn to think differently about code extensibility? Will they learn to think differently about resiliency? Will they learn to collaborate differently, communicate differently, challenge differently? Or will they learn that mistakes result in fewer choice assignments and poorer performance ratings? Will they learn that mistakes make them look bad? Will they learn that mistakes impact their opportunities and career growth?

The way that you as the leader respond to the mistakes goes a long way in influencing the way your team will respond to mistakes. Are you creating a learning environment, or are you paralyzing your team and squashing creativity? Think through this. Find examples. Are you leading the way you would want to be lead? Are you treating someone else's mistakes the way you'd like your leader or your colleagues to treat *your* mistakes? You as the leader need to understand how your team responds to their mistakes, and how *you* respond to their mistakes.

People make mistakes. I make them. You make them. Your customer makes them. Your team makes them. Smart people everywhere make mistakes. You are the leader: what do you do when you know your team has made a mistake? Hamdy Farid gives leaders this advice:

Consider:

"You have to accept it. Mistakes will happen. The teams are moving fast, so we have time to recover fast. The key word here is 'fast.' If the development team makes a mistake in a sprint, well, worst case is that we've only lost two or three weeks. Is it the end of the world? No, not really. You can recover. But if you don't give me a demo for a whole six months and then come to me and say, 'I'll show you this great thing' and it doesn't work, and it's full of mistakes? Well, that's when mistakes become a problem. Mistakes become a problem not because your team has made them, but because they didn't invite their business partners into the dialogue early and often. They didn't demo. They didn't seek feedback. That's when mistakes become a problem."

Karl Betz helped his team move from fear to trust by how he handled a situation where the team shared that they had made a significant mistake. "There was a time when people were afraid to bring forward bad news. They were afraid to say that they were in trouble. They were afraid to ask for help. And that's a drag on the organization and we missed out on opportunities to learn from those mistakes."

The team shared that they were off schedule because of a mistake they'd made weeks earlier. That mistake took them down the wrong path. When they realized the mistake, several weeks had passed. The team was hesitant to share, but then finally the team leader told Karl that they were behind. They redid the estimates and found that they would need another month.

Karl accepted the input and their re-plan. He emphasized how valuable it was for his team to share the facts and a revised plan, and he openly expressed his appreciation. Then he stood up and gave everyone a high-five. "The looks on their faces! Their jaws were on the table. They were shocked that I wasn't yelling and screaming. I told them that I didn't like that we were behind, but that I *did* appreciate that they came forward with the full facts and data, and they came forward with a plan to move forward. They didn't sugar-coat anything, and I celebrated their honesty and transparency."

Karl didn't dwell on the past. He asked how the team was going to course-correct. He asked how he could help. And he emphasized his expectation that his team bring mistakes and misses to him sooner. Not only on this project, but on any project or deliverable.

"Now, if somebody says, 'I made a mistake,' I say, 'Okay, let's rally. Let's figure this out.' Once we've figured it out, we take a step back. I ask the team to identify and discuss lessons learned." The dialogue that Karl had with the team and the way he showed up in the dialogue generated trust. The dialogue generated a feeling of safety and helped to reduce fear.

Challenging them.

Challenging your team starts with knowing what they're good at and knowing what they're interested in. Your approach might be to immerse your direct report in a challenging stretch assignment. Or your approach might be to immerse them in a different role. With new roles, be clear with your expectations the same way you are clear with expectations for existing roles. And perhaps most importantly, give them real work, not simulations.

Hamdy stresses this point.

Consider:

"Rotate them through multiple situations. I'm saying, 'situations' instead of 'positions.' You rotate them so that they start developing an understanding of the entire machine. Whether that's the R&D machine, whether that's the product machine, that's not what matters. What matters is the immersion. They will grow much faster by exposure to different angles of the same problem.

"Let's say I'm developing a feature for a telecommunications company, and I have a talented engineer. Great. That person might be great in developing software but let me put him into the delivery organization that's delivering the feature that he developed. Then later, I can send him to you to talk to you as an actual user of the feature, so that he gets exposed to how people are interacting with his or her feature.

"I focus on getting them exposure to different angles. As they start solving problems, they will have better understanding of the bigger ecosystems, the bigger situation that they have to solve. The best engineer for me is someone who does not shy away from going on site, delivering the product, presenting the product, talking to marketing about the product. Someone who's willing to do that? That's an engineer who is willing to grow."

Early in Harry Harczak's career when he was a young auditor, and later in his career developing talent, he found that what mattered most in his own development and in developing others was *real experiences*. Harry's career included many varied senior leadership roles. Harry J. Harczak Jr., is a former audit partner of Coopers and Lybrand (now Price Waterhouse Coopers), was a senior executive at CDW Corporation where he served in several senior leadership roles including Chief Financial Officer. Next, he was EVP of sales and EVP of marketing, business development, and product management. He shared his view on helping people grow.

Consider:

"A lot of talent development is done through experience. Early in my career, in my formative years, I had the good fortune of working for a lot of different leaders. I was in public accounting, and I got to work on different clients. Each client had a different engagement team. When I'm a young auditor, I'm working for one manager, then when that job's complete, I go to another job working for a different manager and a different partner. So, I got to see a lot of leadership styles. I got to learn first-hand about a lot of different leadership qualities, and then I was able to pick and choose those that I wanted to emulate and those that I wanted to discard. That experience helped me develop.

"The other part of development from my experience—both my own development and people that I developed—were the experiences that you have and that you give early in a career. In my own career, I progressed rapidly. I was given assignments that I never thought I would be doing after two or three years. That was proof for me that the more ability I showed to take on additional assignments and learn on the job, the more opportunity I got to do more and more.

"I'm not against having formal development plans and having people take certain classes, but there's no substitute for learning and developing on the job. You get an assignment, you're not always going to be told how to do something, so you figure it out. You show that you can figure it out.

"Given my experience in my own development and in developing people, there's no substitute for getting and giving stretch assignments, asking people to do more than they think they could do, and providing candid feedback to them."

Usha Arora shared with me her approach for developing talent.

Consider:

"I encourage my team members to go to surrounding areas and learn new and different things. Of course, they must be sure that they are delivering what they are supposed to deliver. But they also have other things that they can do. They can explore. They can learn. Then come back to the team and share their knowledge. They host brainstorming sessions on what they have found interesting, and they share with the team. The key goal is that they are constantly learning and feeling that they are doing something more. They are learning something new, and then they can share that information with the team. Frequently we find that ideas that came up in a knowledge sharing session resulted in us either incorporating certain features in the product or establishing a new process or new tools that increased productivity. It's very positive reinforcement when something that they came up with themselves gets utilized across the board."

Know your team. Know each member of the team. Know how they feel about their role, how they feel about their work, how they feel about the team. Learn and know what each wants from his or her role, from the team, from you as the leader. Learn and know what each wants in terms of his or her career.

For those who prefer to remain in roles like the role they're in today, there is a path that you and your associate define together. There are any number of tools that you can use to help with this: learning maps, experience maps, development plans.

Create a structure to help them organize and control their learning. A structure I've used separates and categorizes the role of the leader into four domains: culture, operations, strategy, and financials. Create learning plans and development assignments—*real* assignments doing *real* work with *real* accountability and outcomes that will be measured and that matter.

Enabling them

Development plans

As the leader, you are responsible and accountable for building talent on your team, both for your team *and* for the broader organization. Developing talent is one of your obligations to the company. You can use the following as a guide to share with the people you're developing.

To be clear, not everyone on your team should be developed the same way. As we've said, each person is an individual with specific strengths and weaknesses. As the leader, you are expected to know the strengths and weaknesses of each person on your team. You then create development plans specific to each. Some of the people on your team will rise to the challenges, and some can't or won't. You need to know which. Then tailor and execute the development plan accordingly. Your plans and your commitment to developing the talent on your team will not be equal across your team, but you must be fair in your commitment to each person and to the opportunities you create for each person.

Let's get started.

Co-create a development plan with each employee.

The employee writes down what they want to do and why they want to do it. They write down the skills they want to acquire and the knowledge to seek to gain. You as the leader bring a leadership perspective and a company perspective to the plan. You challenge the associate—not second-guessing but seeking to understand—on what they want and why they want it. With the insights you gain from the dialogue, you help identify opportunities for the employee to have the development experiences and gain the insights they want. You have broad experience and perspective across the company; you can help identify opportunities with other projects, with other teams, and with other leaders that can help the employee grow the skills and learn the competencies that they aspire to learn. You also have a perspective on the skills that the company needs. Share this insight. Your insight might inspire your employees to learn skills or prepare for a path that they hadn't considered until you made them aware of the opportunity.

Once the development plan is complete, your employee should use it as a roadmap. You both should revisit it at least every 6 months. If your employee is new to their role, they should revisit it more frequently for the first year. Assess progress, adjust, and adapt the plan as necessary. The objective isn't the plan; the plan serves the objective. If the plan isn't serving the learning and development objective, the plan should be modified. Keep it specific, keep it actionable, keep it oriented to the learning and development objectives. Keep it relevant. No one wants to work the wrong plan.

Regarding the importance of keeping objectives in perspective, keep in mind this quote from Winston Churchill. He said, "Success is not final. Failure is not fatal. It is the courage to continue that counts." Keep this in mind. Keep your objectives in perspective. Of course, your objectives matter, or you wouldn't be pursuing them. Of course, your objectives matter, or you wouldn't be committing resources to achieving them. But it is likely that no one will die if you miss the objective. Your objectives are important. They matter. And it is likely that lives do not hang in the balance. Recognize this distinction and behave accordingly.

Another key activity that is often overlooked with development plans: share your plans. Both you as the leader and your employee should share the plan with colleagues and with leaders. Make it known what your employee is working on and how they're working to develop and how they seek to grow. It doesn't serve the employee if no one knows that they aspire to do and how they're planning to get there. Employees are sometimes reluctant to share their aspirations, whether out of fear of seeming too ambitious, or fear of failure, or any number of other fears. Sharing the plan can inspire her colleagues, and it makes leaders in other parts of the company aware of her aspirations. No one succeeds or advances alone, so make people aware. When colleagues and leaders know what she's working on, they can offer suggestions or advice or support. Point is, others can help.

And once the development plan is underway, support her and remind her that every challenge and every stretch assignment is an opportunity to demonstrate that she's dependable, that she honors commitments, and that she keeps her word. Public commitment to a development plan is a powerful motivator to get it done. It's a personal commitment made public. It's accountability. A colleague defined accountability as keeping one's commitments in the eyes of others. Accepting accountability and owning outcomes is critical to an employee's growth as a valuable member of the team.

Set the expectation that your direct reports must deliver results.

Once you're clear on how the people on your team want to grow and once you're clear on what each person aspires to, work together to create opportunities for them to deliver results in their own domain *and*, so that they can demonstrate their versatility, create opportunities for them to deliver across domains. These different opportunities across diverse domains—Marketing, Sales, Corporate Governance, Supply Chain—will provide for unique developmental experiences. Taken together, these critical experiences serve two critical functions: 1. To "round out" the employee, and 2. to enable them to build connections and credibility across the business, outside of their day-to-day work. The types of opportunities you should help create include:

- *Strategic* engagement on critical vendor contracts, with measurable outcomes and benefits. Is there a technology RFP that your employee could lead? Are there upcoming vendor contract negotiations that your employee could participate in and perhaps lead the technology services and deliverables part of the contract?
- *Cross-functional* leadership of complex projects with high visibility. Is there a company-wide initiative to determine optimal approaches for remote work, including such factors as rent vs. buy economics for office space, cultural considerations, and customer requirements and expectations?
- *Financial* fluency and stewardship of budgets across the technical domains. Is there a Value-Based Budgeting exercise that he or she could participate in, or even lead? Is there a project for deriving and measuring unit costs across technology teams that he or she could lead? Is there a company-wide cost-savings initiative that your employee could lead? Is there a customer journey mapping project that your employee could lead?

You as the leader need a system that enables you to identify opportunities and create opportunities for your employees, so they in turn can show their capabilities to their peers, to you as their leader, and to themselves. This is through presentations, conversations, vendor meetings, and many other ways, where your employee gets to show how they think and what they can deliver. This takes a systemic approach to supporting your direct reports by creating such opportunities. This requires that you as the leader relentlessly put the right people in these opportunities to give them the chance to shine.

Giving Feedback

People want to know how they're doing. They also want to know how their team is doing. They want to know where they stand, they want to know their progress, and they want to know that *you* know. It's important to your team to know that they are making progress toward a worthy goal.

It's also important to your team that *you* as the leader know they are making progress. It is important to your team that you can represent their work and share their progress with other people in the organization. Your team wants to know that you understand the value of their work and that you are proud of their work. They want to feel the pride that comes with knowing their work matters. To this end, do more than measure their progress. Measure and then provide visibility to key stakeholders on how your team is doing.

Think about your own behaviors. Are you asking for real feedback, or are you asking for feedback that will affirm your own hopes and that will gloss over your weaknesses and your opportunities? How do you get real feedback? How do you get actionable feedback? Asking for feedback, and not *really* meaning it, is much like greeting people with, "How are you?" "Fine," they say. "How are you?" they ask. "Fine," you say. It was nice of you to ask. It was polite. It was conventional. They answered as you expected (and probably hoped!) they would. And you answered as they expected you would. It was nice. It was polite. It was easy.

If you're going to ask for feedback this way, I would suggest that you not bother asking for feedback. You likely won't hear anything more than a conventional and expected and safe response. "Hey, how was my presentation?" "It was great!" "Thanks!" Really? Instead of asking questions that you hope get answers that affirm your hopes and that don't challenge you, here's a question that I encourage you to start asking right now. You can ask this after you make a presentation. You can ask this after a meeting where there was group dialogue. You can ask this after you provide, or respond to, performance feedback. Here's the question:

"Is there any one thing I could have done differently to be more effective?"

Think about this question. Consider asking this question of a trusted colleague after a meeting you've both been in, where you engaged in a dialogue on a sensitive or complex or strategic topic.

Before you ask anyone this question, think about the way the question is worded. It's very specific: "one thing." You're asking for an assessment of one thing. You're not asking for a lengthy explication of your presentation. You're asking for one thing. The "one thing" part of the question makes your request for feedback much easier to answer. You're helping your colleague narrow and focus. "Done differently." Of course, there are any number of things you could have done differently. You're asking for one thing.

And consider "more effective." You're *not* asking them what you need to do differently. You're asking for a judgment of effectiveness. You're asking for an assessment of degrees of effectiveness. This question, and the way it is worded, makes it much more easily answerable. You create a safe space. You're making it safer and easier for your colleague to answer you and to provide you with input that you can assess and weigh and consider.

Consider this: if a colleague asked you this same question, how does it make you feel? When a colleague asks me for feedback, I reframe their question. I reword it. I ask myself, "Is there any one thing that she has done to be more effective?" This reframing helps me think more incisively and answer more specifically. Reframing the request for feedback helps me provide input that is specific and actionable. It's positively oriented, in that I'm not focusing my response on what I thought didn't go well or what didn't work or what didn't "land." I'm not focused on the negative. Rather, my response is intended to be generative, to take what she said in her presentation or in the dialogue and then reframe it in a way that makes it better: more specific, more precise, more effective, or more persuasive. I'm not telling her what she shouldn't do. Instead, I'm offering her a point of view of what "better" would look like and sound like. It's the difference between "that didn't work" and "this might work better." It's affirming. It demonstrates on an even deeper level that I believe in her, that I believe she can be even better, and, by answering thoughtfully, that I'm making an investment in her future success.

This approach takes practice. It takes practice answering thoughtfully and specifically. It takes practice to orient your feedback in a way that makes it generative. You must listen carefully to respond thoughtfully. It moves giving and receiving feedback from being a binary exchange to a generative exchange. After all, feedback is intended to provide input and guidance for her to consider the next time she presents or the next time she's in an important dialogue. Feedback should do more than evaluate what has already happened. The best feedback is thought-provoking because it is specific and generative. It should do more than help us feel good or feel bad about what we just did. It should prepare us to be even better the next time.

When you are giving feedback, be relentless but not ruthless. Your direct reports must seek feedback, from you and from others.

Anyone who takes their professional development seriously must seek feedback. Anyone who doesn't want to seek feedback, or simply won't, won't develop into the professional that you and your company need them to be. As the leader, you must model the way: show your team that *you* actively seek feedback and that you adjust for the feedback you receive. Your team needs to do the same.

One of the ways to do this is to teach them to orient discussions with their leaders (you, the business unit leaders, leaders outside their domain) from a baseline of "coach me from the baseline of excellent performance, not simply satisfactory or acceptable performance." I received a lot of direct and challenging feedback over the years that didn't make me defensive, rather led me towards purposeful adjustments. Model the way. Show your team what you do and how you do it. Specific opportunities for your team include the following:

- Seek to understand how they're perceived in the workplace. I had one employee, a rising start by anyone's definition, who early on was labeled "ambitious." There's nothing wrong with being ambitious, but if that's the first word that comes into people's minds when she walks into the room, that's a problem. The negative connotations of "ambitious" include "only focused on themselves," "all they want is the next promotion," "she acts like she is better than everyone else." And none of these connotations works in favor of your employee. She needs to seek out and learn how she is perceived. Ambition can drive the way she approaches her work, which is fine, but ambition shouldn't cause her to create an impression among her colleagues that she is more interested in herself than in her contribution to the business. She risks being viewed as upwardly oriented and less focused on working her current job or building her team. Guide her to focus on doing great with the job that she has and building a solid team under her.

- Make tough personnel decisions. If she's leading a team, she needs to demonstrate that she holds her team accountable to high standards of performance and high standards of behavior. Another of my direct reports earlier in my career was leading a small team—15 people—and had one direct report on that team who was perceived as a marginal performer. As the leader of that team, and the direct leader of that person, he was viewed as being unwilling to drive changes, unwilling to hold people accountable, more concerned with getting along than with getting great work done. If colleagues don't have confidence in an employee, then, often, they will also lack confidence in that average performer's leader. Getting this feedback—that they're perceived as unwilling to make tough people decisions—will force your direct report to get clear about his own perspective. This will be instructive for him. If his marginal performer is a diamond in the rough, he needs to help that person grow into his potential. If his marginal performer is no better than a marginal performer, he needs to make a change—either by causing the marginal performer to improve, or by moving him out of the organization.

Your direct reports have opportunities throughout the year and throughout their careers to seek feedback. What matters most is that they actively seek it, deeply engage with it, and adjust when appropriate.

Your direct reports must own their performance.

This means taking accountability for what works when things work and for what doesn't work when things fail. This means owning their actions and the results, impact, and output of those actions. Sometimes projects go well; sometimes they don't. We win some, we lose some. Mistakes and misses happen. Owning them demonstrates a humility and self-confidence that others will see and will respect.

And there's a difference between "mistakes" and "approaches." Mistakes are actions or judgments that turn out to be wrong. Approaches are attempts.

Addressing Failure

A word about failure. There are many ways to look at failure and many ways to respond to. Let's start with a few definitions. From Oxford Languages, the most simple and straightforward definition: "Lack of success." Further: "the omission of expected or required action." From Merriam-Webster: "omission of occurrence or performance." None of these definitions equates failure with the person. Failure is what happened. Too often I've seen employees—young, old, experienced, inexperienced, tenured, new to the job—equate failure with their sense of self. Failure is what happened; failure is not who you are. A failure can be a missed expectation, an output not aligned with the expected output, or the dissatisfactory result of an effort.

I challenge you to define failure not as an endpoint or an outcome. Sometimes things work as we expected, and sometimes they don't. In my view, failure is choosing not to engage. Failure is choosing to avoid, choosing not to try, choosing to give up. That's failure. Wins and losses will happen. T.S. Eliot wrote in the *Four Quartets* that we are undefeated so long as we go on trying (Eliot, 1943). Albert Einstein: "Failure is success in progress." I had a leader early in my career who insisted she had never failed at anything. My response to this when I heard her say it was to think one of two things: she had never tried to do anything that was very difficult, or she wasn't telling the truth. I suspect the latter.

How is this connected to giving your team the environment and support they need? Define for your team how you think about failure. Tell them what you believe failure is. If you believe that failure is the choice not to try, tell them that. If you believe that failure is missing a deadline or a key commitment, tell them that. Be clear in what you believe and why you believe it. Share your experiences. Tell them when you've missed the mark, when you didn't get the desired outcome, when what you created or designed or developed did not meet expectations. And tell your team that they will likely do the same things during their career.

And the cliché, "Fail fast"? I don't like it. Here's why: when I think about this notion practically, who in their right mind sets out to *fail*? Do you? Does your team? I doubt it. What is implied in the phrase "fail fast" is that we try, we act, we do, and then we *learn*. We *learn* from what we tried. We *learn* from what we did. So, let's change "fail fast" to "learn fast."

Creating Incentives

Design incentives to keep the teams aligned, both the product teams and the technical teams. Consider Mike Brendzal's experience. He's seen cases where the product team was driven by a date, and the technical team was driven by their own set of requirements.

Doug Lowell has a solution: "Set up a reward system to favor people who leverage and embody the behaviors that you want. You do want to measure outcomes. You also want to evaluate behaviors. If you're going to live by a system of principles, you want to reward people who follow the principles."

Leaders are deliberate in designing the reward system. You are the leader. Do you want to encourage trying? Do you want to encourage risk-taking? Do you want to reward failure? If you don't, your team won't take risks. If you do, you must establish incentives that encourage these behaviors, and you must follow this with a reward system that rewards your employees for demonstrating these behaviors and demonstrating these actions.

The goals that you set will reveal whether you are encouraging risk or discouraging risk. There are many different types of objectives (for example, SMART objectives, where the objective is Specific, Measurable, Attainable, Relevant, and Time-Bound) and many ways of measuring performance against those objectives (for example, the OKR approach, where you identify Objectives and Key Results). If you want to see your teams take risks, then the goals you set (with their input, of course) need to challenge norms and old notions and perceived constraints (recall Roger Bannister and the 4-minute mile).

Mentoring

Mentoring, like "support," is a big word. It means to teach, to assist, to advise, to train. In practice, there are two dimensions to the mentoring relationship: what the mentor learns and teaches.

First, encourage each employee on your team to seek a mentor for specific reasons. They may want to learn a skill, in which case suggest a teacher rather than a mentor. They may want to talk through the ways you approach software development challenges or the challenges that are part of team dynamics. They may want career counseling. They may want help establishing and growing your professional network. There are any number of reasons to want to work with a mentor. Require that each of your associates gets clear why they want to do this, because they will need to be just as clear when they sit down and begin the work with their mentor. Get clear and write it down.

When it comes to mentoring, your employees must get specific on two key areas. Laurent Therivel, CEO of UScellular, has mentored many employees in his career. He's seen in his experience mentoring people that mentees "they fall into two traps: (1) Not being able to articulate what they care about, and (2) Not crafting goals—both personal and professional—that are realistic, specific, and refined." He encourages every employee to do the important work of determining and deciding what's important. "People need to get clear in their own thinking what they care about. This clarity makes it easier and clearer for them to make decisions about career options—do I take this particular job, do I go after this other particular experience? It's completely reasonable that these choices change over time, but you've got to have a starting point to help guide you and direct you." And being able to articulate this starting point—both what you want and why you want it—will help you and it will help your mentor structure your work together in the most productive way. After all, the mentor-mentee relationship is nothing if not processual. The hard work of making decisions about your career and your goals—this process of being able to articulate with clarity—is what catalyzes the hard processual work and helps generate worthwhile outcomes.

Here's a reason not to choose to work with a mentor: "Because my boss told me to." I've experienced this. I had a junior person in the organization approach me to be his mentor. I didn't agree on the spot; I never do. Instead, I asked him what he was looking for in a mentoring relationship. His answer: "My boss told me I should talk to you." "About what?" I asked. "I don't know. Just that it would be good to have you as my mentor." Was he thoughtlessly doing what he was told? Was he checking a box on his development plan? This went on for another minute or two, at which point it became clear that he didn't know what he wanted from a mentor or why he wanted to meet with one. I chose not to work with him.

You, as the mentee, owe it to yourself first and foremost, and then to your prospective mentor, to be clear on what you are seeking from the relationship and why you're seeking it. What do you want, and why does it matter? You owe it to yourself to be clear. And you owe it to your mentor because you're asking her to commit time and energy to you. It's the respectful thing to do.

Traditional Mentoring

I suggest that your employees find two mentors. They should be in different parts of the company, at different levels, and diverse in background and experience. In traditional mentoring, the employee takes on the role of the student. He or she seeks out a mentor for one or more purposes, for any number of reasons. Perhaps the student wants to learn the technical dimensions of a role. This becomes less mentorship than on-the-job training, but I've seen employees seek out and engage mentors for this reason.

If we elevate the teaching from having the technical know-how, which really becomes more like training than mentoring, if we elevate it from "how to do this technical work" to "how to approach and think about this technical work," then the relationship takes on a mentoring aspect.

Here's an example: a young developer right out of school joins an agile software development team. The team's mission is to create a program to handle thousands of transactions per second, perhaps a real-time rating and charging engine or a billing platform. The coding is the coding, and it is guided and directed by the mission. In this case, if this employee is working with a mentor (again, think "trainer" or "instructor" in this context), he or she will be instructed on how to apply his or her technical skills to solve the problem at hand. Pretty straightforward. And all things considered, I'd argue that this really isn't a mentor/mentee relationship. It's a relationship between a student and a teacher.

But let's change the example. Instead of engaging a teacher or mentor for this kind of help, let's say the employee engages the mentor for help in learning how to think about the challenge beyond simply developing the software. Here, the mentor can challenge the employee to think differently about the problem. In this example, where the employee has been tasked with coding software for thousands of transactions per second, the mentor can challenge the employee to think differently. The mentor can challenge with, "How would you approach this problem if the software you're developing needs to scale to handle millions of transactions per second? What then? What about billions of transactions per second? What *then?* How do you think about the problem *in that case*? How do you approach it?" The employee may ultimately create the same solution, but he or she will have been challenged to think more broadly, to scale their thinking beyond the immediate problem. The solution may be no different, but the intellectual angle of approach will have been tested and challenged.

The mentor in this case, if the mentor is deliberate, will be explicit about the teaching: the mentor is teaching the mentee to think broadly about potentials and capabilities, about scale, and about suitability and extensibility. The employee's original solution for the original problem may prove to be the most appropriate solution, but he or she will begin developing the habit of challenging the potential of the solutions he or she is creating. Problems might expand; can the solution set expand right along with it? Developing this habit of thinking, of awareness of the potential of the problem and the potential of the solution, is an important part of the development of the employee.

Mentorship and advocacy from a highly regarded senior leader helps the mentee get the "inside view" of the issues senior leaders face, how they talk about the business, how they define value, and much more. This insight is critical to the mentee's being able to "fast start" in a leadership role.

Let's take a soft-skill, or human-skill, example. Let's say the project is at stage where the product owner is defining requirements. This session is best when it's dialogue. The product owner shares her requirements, and the developer, in seeking to understand, asks questions. The questions oftentimes will progress from seeking to understand the "what" to seeking to understand the "why": *Why* do you want the customer to see this next screen? *Why* do we want the customer to see this information? *Why* do we want the digital buy-flow constructed and presented in this manner?

This can be dicey for the developer. It depends on the developer's mindset, and it depends on how the questions are asked. Questions that challenge might be interpreted and heard as confrontational. Asked differently, the question will be heard and interpreted differently. Instead of asking, "Why do you want the buy-flow to go from this screen to that screen?", the question can be asked this way: "What do we want the customer to be thinking when they go from this screen to that screen?"

In the first question, the product owner may be put on the defensive, feeling that they have to defend their thinking, maybe even feeling that they're being asked to explain customer sentiment to someone (a developer, of all people!) who doesn't understand customer sentiment half as well as the product owner does. This line of questioning, and this line of listening, can feel confrontational. And the response to confrontation will be defensive.

But asked differently, as in our second example, the response is more likely to be an entry to dialogue between the product owner and the developer. Asked differently, the response and the dialogue can become generative. The original answer may be the final answer, but learning will have occurred along the way, through the dialogue. At the very minimum, the developer should emerge from the dialogue with a better understanding of the product design. Understanding the "why" enables the developer to evolve from being an order-taker to being a co-creator.

This is the type of soft-skill education and learning that can result from a thoughtfully crafted mentorship. This goes beyond "What do I do?" to "How can I best do this?" When the product owner feels that the questions about the buy-flow really are questions about his or her expertise or suitability as an effective product owner, the demo can devolve into defensiveness, with people staking out their ground and working to defend their ground. When the product owner feels that the developer is a co-creator and that the developer wants to learn and deepen their understanding of the product, the question-and-answer session can evolve to a dialogue that deepens the developer's understanding. Shared understanding is critically important to productive and generative collaboration.

Mentoring Tenured Employees

Mentoring is not only for new employees. You as the leader need to consider the value of mentoring with and for more tenured employees. Think about your team. Think about how each could benefit from growth in a particular dimension. Consider connecting your tenured team members with senior executives, even Board members. Why? Because senior leaders and Board members have experiences and points of view that could benefit the rising star or the seasoned leader on your team. Another reason to create mentoring relationships for tenured employees: each person on your team wants and needs to feel that they are learning. They want to know that you take their professional growth seriously and that you remain as invested in it today as you were when they first joined the company. It's positive reinforcement.

Reverse Mentoring

This is part of the written agreement between the mentee and the mentor. The mentor and the mentee "change places." The mentor becomes the student, and the mentee becomes the teacher. The content of the material can be hard-skill related or soft-skill related. Example: the mentee teaches the mentor how data scientists curate data and apply it within certain data models to create specific outcomes. He teaches his mentor how these same data scientists think about the structure of the data models, and why they think that way. The mentee teaches the mentor, who in turn finds ways to help the mentee, who in turn becomes more effective. Attentive and specific reverse mentoring can create a virtuous cycle. This bidirectional learning and teaching relationship is one more way that leaders can inspire and support their technology teams.

Part of the value of reverse mentoring is that it gives the leader the opportunity to learn how the employee thinks. Anyone can see *what* someone does; we gain insight into a person and can begin evaluating them as a potential leader when we understand how they think. This *how* emerges in dialogue.

When Does the Mentor/Mentee Relationship End?

It ends when you've satisfied your objective(s). You'll know this, because on entering the relationship, you documented and shared what you wanted, why you wanted it, and how you would measure your progress to achieving it. This discipline at the beginning of the relationship helps you get clear on what you want and what you need, and how you'll know when you've achieved it. Because you have clear objectives and you're measuring your progress, your status will be clear, and it will be clear when you're done.

Model the Way

Teach your team how to challenge. To be a good teammate, they need to show up. Showing up will sometimes require that you challenge statements or ideas or strategies or directions. Challenge ideas, not people.

How to challenge: challenging implies confrontation. Most people don't like confrontation. It rarely brings out the best in others or in ourselves. One way to challenge a new idea, a new promotion, or a new marketing campaign, is to seek to understand the desired outcomes. Don't ask, "How will you know this is working?" Instead ask, "What will this look like if you roll it out and 2 or 3 months later, it's totally, completely exceeding your expectations? What would that look like? What more would it take in terms of people or money or time to make that happen? And how do you think about what it would look it if it wasn't meeting your expectations? How do you think about that?" These questions are less likely to put the other person on the defensive. These questions are more likely to engage them in dialogue. And here's a key: in the dialogue, you'll hear and learn more about the project or promotion, you'll also hear and learn how that person thinks.

Key Takeaways

1. Listen.
2. Show you care by listening to what people care about.
3. Show you care by listening to what people need.
4. Show you care by creating an environment where people can do great work in service of something bigger than themselves. This includes the physical environment, the social environment, and the professional environment.

References

Eliot, T. (1943). *Four quartets*. Faber and Faber.
Levy, D. (2000). Applications and limitations of complexity theory in organization theory and strategy. *Computer Science*. https://doi.org/10.4324/9781482270259-3
Lewin, A. Y. (1999). Application of complexity theory to organization science. *Organization Science, 10*(3), 215.
Rosenhead, J., Franco, L. A., Grint, K., & Friedland, B. (2019). Complexity theory and leadership practice: A review, a critique, and some recommendations. *The Leadership Quarterly, 30*, 1–25.
Stacey, R. (1996). Emerging strategies for a chaotic environment. *Long Range Planning, 29*(2), 182–189.
Tsoukas, H. (2017). Don't simplify, complexify: From disjunctive to conjunctive theorizing in organization and management studies. *Journal of Management Studies, 54*, 132–153.

13

Agile Principle 6: "The Most Efficient and Effective Method of Conveying Information to and Within a Development Team Is Face-to-Face Conversation"

Abstract

Delivering valuable software frequently is the name of the game. The most efficient and effective method for conveying information is that method—whether in person or virtual, synchronous or asynchronous—that maximizes your three most valuable resources: your time, your energy, and your attention. Efficient and effective communication matters because it is the sole means for causing and ensuring teams stay aligned on their objectives.

Vignettes from leaders at Centene, UScellular, and Oracle illustrate the challenges of efficient and effective communication.

The medium you choose for communicating with your team matters less than the content of your communication. With multiple mediums—face-to-face, video, instant message, email—at your disposal, you get to choose the medium best suited to your purpose. Recognize that with whichever medium you choose, you must be consistent in how you use that medium and what you use it for because with each message you send or dialogue you engage in, you're setting an expectation with your team on how you will use the different mediums and for what purpose.

What Does This Principle Mean?

What does this Principle mean? Here's a way to think about this: "We value face-to-face conversation. We prioritize face-to-face conversations. We value generative connections. We prioritize efficiency. We believe that face-to-face communication is the most effective and efficient way to share information, discuss, and make decisions rapidly. Face-to-face conversations allow for quick shifts, quick prioritization, and quick decisions."

Let's deconstruct this Principle.

"Efficient." In general terms, "efficient" is the maximum output or effect with the minimum required input or effort. For our purposes, "efficient" communication makes the highest and best use of a person's time, energy, and attention. There are many ways to define "efficient." A couple that I like:

- Coursera's five Cs: Clear, Correct, Complete, Concise, Compassionate (Coursera, 2022).
- From www.theinvestorsbook.com: The elements that make communication effective: Clear message, correct message, complete message, precise message, reliability, consideration of the recipient, sender's courtesy.

"Effective." According to a leadership article by Cheryl Keates, PCC, that appeared in the online edition of Forbes on September 10, 2018, the five Cs of effective communication are: be clear, be concise, provide a compelling request, be curious, and be compassionate (Keates, 2018). When you are clear, you are speaking in a language that you and your listener both understand. This might seem obvious, but I've seen leaders fall into the trap of using language that is particular to their domain, using a vernacular, that may not be as clear to the listener. Remember, effective communication is *connection*, it is not *transmission*.

Coursera too has a great description: "Effective communication is the process of exchanging ideas, thoughts, opinions, knowledge, and data so that the message is received and understood with clarity and purpose. When we communicate effectively, both the sender and receiver feel satisfied" (Coursera, 2022).

"Conveying information." To "convey" is to carry or to transport. It is active. In this case, the payload being carried or transported is information.

"Face-to-face conversation." In our day of multimodal communication, distinctions become increasingly important. A conversation that is face to face implies a verbal, real-time, synchronous exchange. Strictly speaking, "face to face" is in person, eye to eye communication. That's the definition. Face-to-face conversations are well and good when teams are local and are in the office every day. But what about today's environment, where development teams are distributed around the country or around the globe, and where teammates meet in person infrequently or maybe not at all? What then? In practice, the medium for this exchange is multivariate. More on this below.

You Are the Leader. What Do You Do?

Start with the End in Mind

What will this look like when you've achieved it? When this Principle is in place, your team will be sharing information about pressing work in real-time, addressing pressing topics directly, clearly, and efficiently.

Listen and Learn from Others

How do you get to the "end in mind"? You start by asking great questions, and then you listen.

1. What is "effective communication"?
2. What is "efficient communication"?
3. How will we know when we're "efficient"?
4. How will we know when we're "effective"?
5. How can I help?

We have deconstructed this Principle to get at its meaning. We have defined the words, and we are clear on what the words mean. We have defined the objective with this Principle by starting with the end in mind. We have asked great questions and listened, really *listened*, to what we were told. Now it's time to learn from others.

What Is Effective Communication?

Verbally, communication is the conveyance of information from one person to another. The *understanding* of the information, the *interpretation* of the information, depends to a significant degree on the way in which it's conveyed, including the body language, the facial expressions, the posture, gestures, and tone. "How much of communication is nonverbal? It may not be exactly 90%, but nonverbal communication—eye contact, smiling, hand gestures—heavily influences how people interpret and react to information" (The University of Texas Permian Basin, 2022).

What Is Efficient Communication?

Communication needs to be both effective and efficient. Communication that is effective but not efficient is a waste of one or more of the three resources at your disposal: your time, your energy, and your attention. Usha Arora shared her perspective on this:

Consider:

"I feel that it's very important to bring the whole team together from time to time. The whole team—whatever role they are playing, whether they are product owners or they're developers or they are providing some domain expertise in general—bring them together and share overall the progress and the challenges, share where we are going, share how we are going to change, share how we are doing. It's very important. Communicate with your team and build relationships with your team."

Is face-to-face communication *always* the best approach? Are there moments or times or topics that are better served differently? This may be where we need to deliberately restrict the application of this principle and take it literally: "the most efficient and effective method of conveying information to and within a development team is face-to-face conversation."

Consider the work of researchers Bloom, Han, and Liang in their 2022 working paper (revised 2023), "How hybrid working from home works out." Their current work sheds light on what constitutes efficient and effective communication in our current hybrid work model. They shared a conclusion from interviews with employees that these employees "became accustomed to a more asynchronous written style of communication, carrying this over to their days working in the office" (Bloom et al., 2023).

Consider:

"Employees reported in discussions that if they had to ask a simple question about coding, a product, or a customer, they were now more likely to do this by message rather than in person...This increase in messaging by ... employees happened rapidly, with no trend across the experiment. Hence, the change in behavior was immediate and persistent (p. 13).

"We see, first, that ... employees both send and receive more messages overall, and second, that this is particularly between employee pairs. This highlights this change in communications, in that ... employees became more comfortable messaging on their work from home days, carrying this over into office days (p. 13).

"Working from home leads to increased messaging both at home and in the office, particularly between [employee] pairs. This increase in messaging is greatest for team members and existing close contacts suggesting hybrid-WFH may lead to some mild silo-effect for individuals, highlighting the importance of the office days for employees to network and build up weaker ties" (Bloom, Han, & Liang, How hybrid working from home works out, 2023).

Here's an argument that finds this Principle troubling.

Consider:

"I was troubled by Principle 6 because it seemed like the objective was efficient and effective, and what it left out to me was productive but also satisfying. It didn't take the worker's perspective. So, what if the principal led with the most satisfying method of conveying information to and within a development team? Satisfaction and fulfillment to the worker is in my view what drives productivity and creativity and efficiency.

"I disagree with the conclusion that face-to-face conversation is always the best mode. I think that is very personality specific and for some people that's true. You can have an initial face to face conversation when you're in goal-setting mode or brainstorming mode, but then you let them go off and work without a lot of interaction. Face to face may not be best. Some of your team might feel like you're meddling.

"An example of that in the real world was the development of Java, the software system. Java was an open-source software technology that was developed by whoever wanted to develop it. It was a team that never met face to face. There were Java conferences, but most people never came. Most of their work was done virtually." (Interview with Doug Lowell).

When we consider "efficient and effective," do we also consider whether that conversation and that communication is "satisfying"? Do we consider whether it is "fulfilling"? The principle does not include these considerations explicitly. But can you make the case that when the purpose of the communication is to "convey information," that "efficient and effective" do need to be "satisfying"? And if the target or the audience of the communication is the development team and *only* the development team, does the communication *need* to be anything *other* than "efficient and effective"? You need to ask your team this question, and you need to listen to their answer.

Are You Ready?

You started by deconstructing this Principle to get at its meaning. You have defined the words and you understand the words. You've defined your objective with this Principle by starting with the end in mind. You've defined what matters most. You know why this matters most because you've asked great questions of your team, your boss, and your customers. You've asked great questions and listened, really *listened*, to what you were told, and you have learned from their stories.

Now that you are clear on what it means to communicate with your team efficiently and effectively, what do you do? You are the leader. It is time to act.

Do

Stay true to the purpose of the communication. Focus less on the medium for the communication. Remember, this principle was written in 2001, before Teams, before Zoom, before Webex. Tools can and should be used to complement, and in some cases replace, face-to-face conversations for sharing information on status and workflow. This principle emphasizes the sharing of information. If your *primary* purpose is to *convey* information, the spirit of Principle 6 is suitable. The Principle is specific to "conveying information." Is there an application of this Principle *beyond* the conveying of information? If we remain true to the strict definition of conveying information, then direct and immediate communication may be the most efficient and effective means of communicating. But can this principle be applied in situations where the objective isn't restricted to conveying information?

We've talked about face-to-face communication, whether it be an in-person meeting or a virtual connection. What about group meetings? Are these better, more effective, more generative, when the meetings are face-to-face and in person versus virtual? The answer is that like most everything else, it depends. It depends on the type of meeting, it depends on the intended outcome of the meeting, it depends on the length of the meeting, it depends on the number of people in the meeting, and it depends on the relationships among the people in the meeting.

One benefit of holding group meetings in person: informal connections. When there's a break in the agenda and people are given 10 or 15 min to get up from their seats and stretch and move around, there is an opportunity for them to connect with each other informally. Most people tend to connect with one or more of their colleagues when there is a break in the meeting. When they do connect, they'll do one of several things: they will chat informally, or they will raise an issue that was discussed in the meeting, or they will strategize how to engage in the next topic on the agenda, or they will follow up on points made earlier in the meeting. When they do any one or more of these, they are connecting on a different level, and they create opportunities to generate new connections and new ideas. In doing this, they are adding or strengthening a dimension to the relationship.

And during the face-to-face group meeting, there are the notorious sidebars. We've all experienced these. We're sitting in a meeting, we're listening, we're focused on the presentation or the discussion. But two people sitting next to each other are whispering or, worse, talking in low tones. They're having a conversation. It goes for 10 seconds, 15, 30, a full minute or more. It's a distraction. If *you* are the one in the sidebar conversation, it is your responsibility as a participant and as a teammate to ensure that what you're doing in your sidebar is generative and not distracting. But better not to distract your teammates with your sidebar conversation. Better not to show disrespect to the speaker or the presenter. Engage in the meeting.

Something else about face-to-face communication: Several dynamics often come into play that are difficult, if not impossible, to experience in video meetings. Consider, for example, the excitement that permeates the room when the sales team shares results that beat the forecast. Or when the leader of the human resources organization announces that the company has earned a "Best Place to Work" distinction. Or when the leader of the finance team shares in a monthly business review that revenue and operating cash flow beat expectations. Another example: when a new person joins the team or the joins the company, the warm welcome simply isn't the same over video as it is in person. Other experiences that feel diluted on video: momentum, anxiety, and tension. These can and do exist in video meetings, but you simply cannot *feel* them on video like you do when you are face to face.

Hybrid or virtual meetings are suboptimal for workshops and innovation. In these cases, in-person meetings are better. It isn't *by definition* better. In-person meetings require purposeful work to leverage the in-person experience and leverage people being in person.

It's the *nature* of the face-to-face conversation that you need to preserve. Consistent with one of the tenets of complexity theory, it is critical in today's environment that you provide and encourage open communication. In today's environment of hybrid work/office arrangements and virtual teams, the nature of the face-to-face conversation can be preserved through technologies like Teams and Zoom. Current technologies enable synchronous connections and synchronous dialogue and synchronous conversations.

You as the leader will conduct various kinds of meetings. Same for your team. There will be daily standup meetings, collaborative backlog grooming sessions, sprint planning meetings, frequent demos, and team meetings. What do you do in each of these meetings? What do you talk about? And what do you encourage and expect your teams to talk about? How do you expect them to show up? Communication can only be considered effective when you connect. Know your team and choose your medium.

Key Takeaways

1. Listen.
2. Synchronous communication can be more generative than asynchronous communication.
3. Effective communication requires a connection—in person or virtual—between you and your team.
4. Efficient communication maximizes the benefits of the medium you choose to communicate in.

References

Bloom, N., Han, R., & Liang, J. (2023). *How hybrid working from home works out.* National Bureau of Economic Research Working Paper Series. https://doi.org/10.3386/w30292

Coursera. (2022). Retrieved from Coursera: www.coursera.org

Keates, C. (2018, September 10). *The five Cs of effective communication.* Retrieved from Forbes: www.forbes.com

The University of Texas Permian Basin. (2022, July 19). *How much of communication is nonverbal?* Retrieved from The University of Texas Permian Basin: https://online.utpb.edu/about-us/article/communications/how-much-of-communication-is-nonverbal

Agile Principle 7: "Working Software Is the Primary Measure of Progress"

Abstract

Delivering valuable software frequently is the name of the game. As with the word "valuable," the word "progress" means different things to different people. Developers define working software as software that functions. Customers define working software as software that has value. Progress toward valuable software is the only progress that matters to your customer.

Vignettes from leaders at Centene, UScellular, Nokia, Amazon, and Oracle illustrate the challenge of defining what working software is, and how to apply the definition of working software in measuring progress.

The most effective, efficient, and productive way for developers and business owners to stay in synch on the development of the product is through demos. Demos properly conducted elicit the feedback from business owners that influence the next sprints. Developers need to listen to the feedback, evaluate the feedback, determine the disposition of the feedback, and communicate decisions with the business owners who provided that feedback. The feedback loop is the generative aspect of the demo process.

What Does This Principle Mean?

What does this mean? Here's a way to think about this: "What gets done" is what we care about, and it is what our customer cares about. We don't let the perfect impede progress. We don't let the perfect become the enemy of good enough. Working software is what we measure and report on. It's the measure of our progress.

Let's deconstruct this Principle.

"Working software." What does that mean? Consider a watch: the hands move, the date changes, but it doesn't keep accurate time. Would you say it works? The hands move. The date changes. Is that considered "working"? It depends on how you define it. It depends on the *requirements* and on the customer's definition of value. If the requirements are for the hands to move and the date to change, the watch works. If the requirement is to reflect accurate, even precise time, it doesn't work. And who should define "working"? The watchmaker? Or the person who needs to know what time it is? Does the software do things, or does it do what you need it to do? Does it function, or is it valuable to the customer? The watchmaker will say the watch works. The person wearing the watch and relying on the watch will say that it doesn't.

"Primary measure of progress." There are (at least) two valid measures of progress: one is from the point of view of the developer, and the other is from the point of view of the business owner. More on each of these below.

You Are the Leader. What Do You Do?

Start with the End in Mind

What will this look like when you've achieved it? When this Principle is in place, your team will be doing developing and delivering software that works. This will be in the form of Minimum Viable (or Valuable) Products and Minimum Viable Features. Your team will be delivering software that works and software that is useful in the eyes of your customer.

Listen and Learn from Others

How do you get to the "end in mind"? You start by asking great questions, and then you listen.

1. What do you mean by "working software"? Ask your team.
2. What do you mean by "working software"? Ask your business owner.
3. What do you mean by "progress"?
4. What is *your* primary measure of progress? Ask your team.
5. What is *your* primary measure of progress? Ask your business owner.
6. What is *your* primary measure of progress? Ask your boss.
7. How can I help?

You've started by deconstructing this Principle to get at its meaning. You have defined the words and you understand the words. You've defined your objective with this Principle by starting with the end in mind. You've asked great questions and listened, really *listened*, to what you were told. Now it's time to learn from others.

What Is "Working Software"?
Working software is software that delivers value in the eyes of your customer. Working software is the primary measure of progress. It is *primary*, because

without working software, you have nothing. But this criterion of progress—working software—is followed very closely by working software that the customer values. The software might function and execute, but if the customer doesn't value it, you have working software with no utility. And if there's no utility, there's no value. Consider the case of a company that did not want to be identified in this book. The CIO wanted to get a lot done and didn't want to be slowed down. He wanted to be first to deliver the new system, and he wanted to impress his boss. IT wanted to move fast, so they wouldn't allow the business team to slow them down. But changing requirements would slow them down. Resolving defects would slow them down. IT wanted to move fast, and the CIO wanted to be first. He wouldn't tolerate delays to the schedule. If the software worked, the software worked, and he could meet his commitment to deliver working software on the date he committed to. But that iteration of the software was not valuable to his customer. Did he meet his commitment?

Heather Ackenhusen related her experience at Amazon, and how she thinks about working software and valuable software. "There is a lot of working software at Amazon that got built and that never got used." But did the team make progress? Yes. They learned what didn't work, what wasn't useful or valuable. What Thomas Edison said about failure can be applied to the notion of working software: "I have not failed. I've just found 10,000 ways that won't work." Again, from Heather:

Consider:

"At Amazon, the teams could produce and demonstrate and deliver working software in increments. The approach was, 'Okay, I've got this little thing of value. I'm going to give it to the customer and then I'm going to build off it and build off it and build and build and build.' Get the working software out the door in a good, usable state. It had to be valuable, but it didn't have to be done and dusted. From the team's perspective, they get to learn from working software. From the customer's perspective, the primary measure of progress isn't that the software works, but that it is valuable."

Your agile software development team may need to iterate tens or dozens or hundreds of times, producing working software at every stage, before the working software becomes valuable software in the eyes of the only stakeholder that matters: the customer.

Making Progress

Progress is, by definition, movement toward a destination. Your agile software development team does the development, and your customer defines the destination.

Making Progress, According to a Developer

Hamdy Farid has been on both sides. He's been an engineer and developer, and he's been a business owner. "As an engineer, progress for me is I have my test driver running. I'm building functionality and building the test driver. I'm simulating data going in and data going out. That's progress for me. The result is of zero value to anybody else. But this is a milestone. I finished the functionality, I did testing, I automated the testing. That's progress. Then I go to the next one. Then I go to the next one."

Take the time to plan. Frame the problem you are working to solve. In fact, one of the biggest challenges can be in the framing. You are the leader. As the leader, you are responsible for ensuring the problem is clearly understood before work begins.

Despite the importance of framing and planning, Dinakar still sees leaders who fail to plan, and teams that begin their work without clear direction. "We still see that sometimes people are in a rush to start sprinting. They don't really take the time to plan. And that results in a lot of rework, which is frustrating."

You and your team are on the hook for delivering valuable software quickly. In addition to lack of clarity, what other frustrations do you as the leader need to be aware of? Dinakar shared that what frustrates his team is when the speed of the sprint results in compromising testing and automation.

Consider:

"Sometimes because of the time constraint that the team is under, they feel like, 'I have to finish this sprint, I have to write code.' So, the things that get compromised are either the number of test cases that they author or the number they automate. Sometimes they author 100 test cases, but they'll automate only 50. So, they end up short-changing that aspect, which is costly in the long run.

"Developers, being developers, sometimes think that code complete is the most important thing. As their leader, I shout from the rooftops that when we say code complete, it means code complete. When we say, 'code complete,' ideally it means that our test cases were run and passed 100%. That's when we say code is complete, not when you finished writing code.

"The demands of speed and showing progress can result in compromising the design. If you get behind early, you risk compromising testing and automation. If the design doesn't get finished on time, it eats into engineering time and the developers' time is compressed, which in turn causes them to use time that should be spent on creating automations and on testing. Ultimately, this compromises quality."

Couple the demands for speed and quality, and something might have to give. Rather than concede the quality that will impact the product that you put in the hands of your customer, provide your team with flexibility. Dinakar has faced this challenge.

Consider:

"We'll give the team a little wiggle room, saying, 'OK, we want you to automate 80% of the tests that you authored because we understand that sometimes there are tests that you cannot automate because of constraints in the tool. That's OK. But I know that sometimes you're running out of time. So, we'll give you 100 days later to complete and catch up on automation.' So, we've done that, we've given those kinds of guidelines. Other times we've said, 'Okay, we know you're in Release One. You should automate 80% of your test cases. Of the remaining 20%, let's say 5% are not automatable. So, the remaining 15%, you need to catch up in the next release.' Even though the team has moved on to other work, you need to require that they hit the new targets by specific dates. We need to track that. This way, we don't build a backlog of test cases that aren't automated."

This is from the perspective of the agile software development team: "The code is good enough because the software works." But what about from the perspective of the business owner? What does the business owner think of "working software"?

Making Progress, According to a Business Owner
In Hamdy Farid's experience as the business owner, he sees and can appreciate the challenge from multiple points of view.

Consider:

"As a business owner, you wear the hat of an actual user of the feature and the capability. From the business owner point of view, progress isn't Minimal Viable Product. Progress is actually Minimum Demo-able Product. A Minimum Demo-able Product is a product that the developer can put in front of me and say, 'Look, this is a capability that we built for you. Some of these buttons do not work, I know, but this is what it will look like. This is how your user journey will look.' It's not 100% completed, but it's progress. The developer showed me something so I can give them feedback. So, to me as the business owner, progress is when the developer creates a demo-able deliverable that I can provide feedback on. For example, say the developer creates some wire frame, some screenshots, some edited graphs and pictures, that's progress. This is one of the beauties of agile."

Leverage the iterative nature of agile software development to make changes as you receive feedback. "One of the fundamental premises of CI (Continuous Integration) is that frequent small builds are best. The goal of these small builds and associated automated testing is to find any coding problems that need to be fixed as quickly as possible. They key to this objective…is to get a full build with as much testing as possible on the smallest number of changes" (Gruver et al., 2012). Hamdy explained that progress gives him something that he can provide feedback on to the developers. The developers in turn take the feedback, assess the feedback, and apply the feedback in ways that improve the product. That's progress.

Are You Ready?

You started by deconstructing this Principle to get at its meaning. You have defined the words, and you understand the words. You've defined your objective with this Principle by starting with the end in mind. You've defined what matters most. You know why this matters most because you've asked great questions of your team, your boss, and your customers. You've asked great questions and listened, really *listened*, to what you were told, and you have learned from their stories.

Now that you are clear on what it means to develop and deliver working software, and how it is the primary of progress, what do you do? You are the leader. It is time to act.

Do

You are the leader. Part of your job in the early stages is to ensure that you have established appropriate measures and metrics that will help you determine whether your team is making progress. Without these, your team will work and work and work, hours and hours, days, weeks. Thousands of hours of effort, but effort that has done what? What is the effort causing? What is it resulting in? Is the effort creating something valuable? How quickly can the effort deliver something valuable? Your business owner and your customers respect that you and your team work hard. But what they need from you is working software. Your business owner can't take to her customer a report that shows how many hours your team has logged. Her customer won't care. Her customer needs something that works.

Remember, in the end, your definition and your team's definition of progress only matters if it's progress in the eyes of your customer.

Key Takeaways

1. Listen.
2. "Working" is in the eye of the customer.
3. "Value" is in the eye of the customer.
4. "Progress" is in the eye of the customer.

Reference

Gruver, G., Young, M., & Fulghum, P. (2012). *A practical approach to large-scale agile development: How HP transformed Laser Jet Future Smart firmware*. Addison-Wesley Professional.

Agile Principle 8: "Agile Processes Promote Sustainable Development. The Sponsors, Developers, and Users Should Be Able to Maintain a Constant Pace Indefinitely"

Abstract

Delivering valuable software frequently is the name of the game. Process matters because done right and done well, it generates valuable outcomes. A constant pace requires high levels of effective communication among all the key participants. Where there is speed, there will be risk. You as the leader cannot allow the pace to create unacceptable risk.

Vignettes from leaders at Oracle and Nokia illustrate the challenge of sustainable development delivered at a constant pace.

Sustainable development at a pace sufficient to satisfy your customer comes with risk. Requirements, and the problems that the team is tasked with solving, must be carefully and clearly framed, or you risk wasting time and effort. There is risk that your development team will sacrifice quality for speed. Progress toward a valuable goal is different than progress for the sake of progress.

Leaders overcome these challenges by framing the problems clearly at the beginning of the development process and ensuring that the development team and the business owner remain aligned on making valuable progress.

What Does This Principle Mean?

What does this mean? Here's a way to think about this: "We like to move fast, and we like to keep moving."

Let's deconstruct this Principle.

"Agile processes." You as the leader of an agile software development team, and each of the developers on your team, know well the processes of agile that we call the software development lifecycle. Leaders and developers alike agree on the process, though they'll name the steps differently, or they will break out one or two steps into three or four steps, higher-level views, or lower-level views. Generally, the steps are these:

1. Planning stage
2. Analysis stage
3. Design stage
4. Development stage
5. Testing stage
6. Delivery stage
7. Maintenance stage

I won't describe each. You can find descriptions from a variety of sources. Later in this chapter, I will address what you as the leader need to do in each stage.

"Sustainable development." Sustainable development means development that your team performs, performed in a manner that is consistent and repeatable over time. It depends on your team, and it depends on inputs that will be delivered to your team.

Your team's mental, emotional, and physical state factor in to creating sustainable development:

- Mental: This includes confidence, incisiveness, achievement-orientation, optimism instead of defeatism, leaning in versus sitting back, and causing something to happen versus waiting for something to happen.
- Emotional: Feeling on the brink of disappointing versus feeling like you're satisfying someone, like you're making good, you're keeping your word. Not only on your team but with your business partners, business owners, and stakeholders.
- Physical: Sustainable pace, sustainable intensity. Avoid exhaustion, physical, mental, emotional, or spiritual.

The inputs that your team will require include:

- Clear requirements defined consistently by the business owner
- Funding approved by the CFO and the finance organization
- Support and direction from senior leadership
- And at the highest levels, an external marketplace that creates demand

"Sponsors, developers, and users." What does this mean? Who are these people?

This is the broader team. Think of this in terms of business owners, product owners, developers (your team), and users (customers, either internal or external, or in some cases, both).

"Maintain a constant pace indefinitely." What does this mean? This means ongoing, iterative work by all the teams involved in the process. The process is interactive and generative. This means that the teams are applying Principle 4, "Businesspeople and developers must work together daily throughout the project."

This means that the teams have *developed* the discipline that agile development requires and that they are *executing* with this discipline.

You Are the Leader. What Do You Do?

Start with the End in Mind

What will this look like when you've achieved it? When this Principle is in place, your team will be doing what they said they would do. They will be delivering on their promise to deliver. They will be delivering at consistent, predictable, and agreed-upon intervals.

Listen and Learn from Others

How do you get to the "end in mind"? You start by asking great questions, and then you listen.

1. What are the agile processes best suited to my development team and my customer?
2. What is "sustainable development"?
3. Why does "sustainable development" matter?
4. How can I help?

You've started by deconstructing this Principle to get at its meaning. You have defined the words and you understand the words. You've defined your objective with this Principle by starting with the end in mind. You've asked great questions and listened, really *listened*, to what you were told. Now it's time to learn from others.

You connect and generate and create and demonstrate. You listen and adjust and deliver and review and deploy, over and over and over again, at a constant pace. But what exactly is "sustainable development"? What is a sustainable pace, why does it matter, and just how important is it really? Speed matters, and a regular and sustainable pace matters so that you and your team can deliver on the expectations that you've set. At the same time, be clear with your team and with your customer about the relative importance and the interconnectedness of pace, quality, and value of the output.

When the agile software development process runs smoothly, it is a thing of a beauty. When your agile software development team and your business owner are in synch, meeting regularly, communicating, collaborating, having generative dialogue, sharing, improving, developing, delivering—when all these things are happening according to your plan, it is a marvel. And it isn't an anomaly or an exception, rather it becomes the standard by which you and your business owner will measure all development teams and all development processes. You've heard the clichés: success breeds success, everyone wants to play on a winning team, and so on.

But you and I both know that software development is as much art as science, and like your investment portfolio, past success is no guarantee of future success. You and your team know how to succeed. You and your team know how to engage with your business owners. You and your team know how to develop, and you know how to develop.

And you and I both know that sometimes things don't go well. Sometimes the process isn't as smooth as it's been in prior development efforts. Sometimes your business owner is less than satisfied with the process. Sometimes he gets frustrated with the process, the deliverable, with your team, or with you. And sometimes your team will become frustrated, too. Agile development is not a panacea. What are the sources of frustration?

There Are Risks

You as the leader of an agile software development team face risks when you establish the framing and set the pace at the expense of taking the time to establish an extensible framework and an extensible reference architecture. Dinakar shared his experience with "teams that tend to think of the overall architecture and solution in small pieces, which is not a good idea because it's very important to have a vision of the final product. The team needs a view not only of the capabilities that are going to exist, but also, what's the technical solution architecture? How are you going to both functionally and technically build this correctly? You as the leader need first to lay out "the end game. You need to first lay that out and then the actual detailed design and development can be done in chunks iteratively."

Agile development teams can be their own worst enemy. Dinakar shared that "because of agile and kind of the misconceptions around it, people sometimes don't spend the initial time bootstrapping to arrive at that overall architecture for the end game." He's seen teams fall into this trap of their own making when they don't think through the end state. The team was busy making progress and then they hit a roadblock. "For the work that we have to do in the sixth sprint, they learned that the foundation was not good enough. They thought that even though they were building in chunks, they shouldn't have to go back and revisit the fundamental foundation."

What do you and your team do to prevent this late-game recognition, after so much work has been done, time has passed, and your business owner is expecting results?

At Oracle, the process is called "bootstrapping." Dinakar describes what "bootstrapping" means to his team.

Consider:

"You take two months where the architects lay out the overall high level architectural solution." This gives teams the chance to understand the reference architecture and helps enforce a development discipline so that people don't approach their point solutions in a disjointed or disconnected manner. As Dinakar aptly summarizes, "I've got to get the foundation right, otherwise it's going to be messy. You have to do your due diligence up front. You've got to take the time and then you can build the detailed capabilities in an agile fashion. That was a big realization."

"Of course, not everyone agrees with that approach. Not everyone agrees that developers need to anchor to a reference architecture." But his years of experience in agile software development have taught him that you and your development team "need a good vision of the end goal before the team starts sprinting."

The Risk of Constant Pace

Dinakar speaks to what he and his team have experienced as the most frustrating and least desirable aspects of agile.

Consider:

"The fact that some teams don't have an overall understanding of what the end game is is frustrating. It's like they're running blind to some extent, and they don't know what's coming in the next sprint. The team is just focused on the sprint that they're currently working on. It's like you're carving a statue, but you don't know what it's going to look like."

Who's to blame? Is it the development team at fault? Is it the business owner? Or is it you as the leader? Dinakar shares his perspective on the role of the leader in this case. "Leadership has to explain the iterative development process so that the customer understands why they are getting a solution in chunks instead of the complete statue with that first delivery. Leadership needs to explain to the customer that his agile software development team is going to deliver in an incremental fashion."

A constant pace is great if you and your team are hand-in-hand and in lockstep with your business owner. Alignment matters, but what makes alignment valuable and generative is alignment *in the right direction*. Heed this exchange between Alice and the Cheshire Cat in *Alice in Wonderland*:

> Alice: "Would you tell me, please, which way I ought to go from here?"
> "That depends a good deal on where you want to get to," said the Cat."
> "I don't much care where—" said Alice.
> "Then it doesn't matter which way you go," said the Cat.
> "—so long as I get SOMEWHERE," Alice added as an explanation.
> "Oh, you're sure to do that," said the Cat, "if you only walk long enough." (Carroll, 1893)

Framing Is Everything

You must start at the beginning, and the beginning is the point where the goal is defined. Approach framing the development project the way you would approach framing a house or framing a good story. This is where a good architecture comes in. You need an architecture that defines the structure of the house, the story, or, in this case, the system. The architecture demonstrates how the elements of the system or the product will interact. Again, think of building a house or writing a story. The architecture defines how you get from one room to the next, from the first floor to the second, from the family room to the patio. In the case of a story, the framing and the structure determine how you get from one element in the story, or from one chapter or even one paragraph to the next. Same with your product architecture.

You as the leader need to ensure that the solution your team is creating—the platform or the system or the product—is framed and architecture appropriately. (More on framing and good design in Principle 9.)

Exiting the Design Work and Entering the Development Work

You are the leader, and you identify these challenges and determine what to do about them. How do you know you're ready to begin developing? Is your team aware of the architecture? Does your team understand how its creations and its solutions will fit into the bigger picture, into the product itself? Dinakar shared that he and his organization have established formal exit criteria for transitioning from design to development. In his experience, what is required for this transition depends in large part on whether his team is working on a minor enhancement or a major change. With a minor enhancement, it becomes less important to spend the time and effort on bootstrapping and on creating technical architecture documents. "But if it's a substantially big feature, if you're creating a new product or a new module, we have a gateway. You create what we call an FSA and a TSA, a functional solution architecture and a technical solution architecture. These two documents have to be created by your team and then reviewed by a team of reviewers. Once it's approved, then you start scripting."

Here again some leaders and some teams will object. Creating documentation can seem at cross purposes to agile development. Dinakar has seen teams struggle with the requirements that they create documentation.

Consider:

"Sometimes it's a challenge because people believe that 'Wait, the whole idea of agile is, you don't have a lot of emphasis on documentation.' Instead of producing documentation, they will say, 'We just discuss this and share through word of mouth and that way we move fast.' But again, when we're dealing with complex development, we cannot take those shortcuts. It may work for simple products. But not for complex projects. We believe it's important to do the due diligence. It's OK if people think it's an overhead. Let's have these documents reviewed and approved, and then we start sprinting."

Doing

It is less important that you shorten your team's development efforts into one-week or two-week or month-long sprints. It is more important that you *sprint*. Gruver, Young, and Fulghum: "Our experience is that the most important thing is to have a regular cadence" (Gruver et al., 2012).

Dinakar shared his practical experience in determining the duration of each sprint. "If you do one-week sprints, it's tough to really can't get anything done because you are getting designs ready, coding it, testing it, writing the QA plans, creating automation for the test plans, and running through all of those, trying to hit 90%+ success rate, and then checking the code. There's no way you can do this in a week. So, we started monthly sprints. We told our teams, 'The design should be ready by this date. Coding should be done by this date. Test plans should be ready by this date. Hand off the development and the demo to product management will happen on this date.' We laid out the whole schedule. We started off like that." His team worked with monthly sprints for a couple of years, and with the experience they've gained, they've cut their monthly sprints down to biweekly sprints.

Whether you and your team structure your work in weekly, biweekly, or monthly sprints, it is critical that you as the leader create a cadence, hold your team to it, and share this commitment with your business owners. In the end, as Dinakar says, "The point of agile is the work that you finish at the end of a Sprint should be deployable. That's the whole point."

Are You Ready?

You started by deconstructing this Principle to get at its meaning. You have defined the words and you understand the words. You've defined your objective with this Principle by starting with the end in mind. You've defined what matters most. You know why this matters most because you've asked great questions of your team, your boss, and your customers. You've asked great questions and listened, really *listened*, to what you were told, and you have learned from their stories.

Now that you are clear on how agile processes promote sustainable development, and what it means for the sponsors, developers, and user to be able to maintain a constant pace indefinitely, what do you do? You are the leader. It is time to act.

Do

Leadership in the Planning Stage
You as the leader bring the right people into the room. You don't bring *everyone* into the room. You identify the *right* people. Who are the right people? Who do you *need* in the room? Remember the distinction that we made between "need" and "want." You may want more people, and more people may want to join. But you as the leader must limit the number of people in the room and the number of voices that can contribute. The process of selecting who is in the room should be straightforward: You need the product owner. You need developers. You need the customer. The product owner secures the funding and shares the objective. The developers bring the technical know-how. The customer defines value. And each group—the product owner, the developers, and the customer—knows the others well enough to understand what matters to them and what they care about most.

Teams that know each other *really well* will also understand the constraints that their colleagues are operating under. Teams that know each other *really well* will understand their colleagues' needs, wants, pressures, and constraints. The team in the room is a republic, representing the voices and the interests of their respective stakeholders. It isn't a democracy, where every person in the organization has a say and gets a vote. No better way to grind progress to a halt than to take the time to hear and respond to every perspective. It's unwieldy and impractical. The people in the room are there by design and on purpose: they have credibility, they understand their own role and the roles of the other stakeholders in the room, and they are committed to progress. You as the leader are accountable for making this happen.

What allows for agile processes to promote sustainable development? The team. The people working together. This is closely linked with Principle 4, "Businesspeople and developers must work together daily throughout the project." The interconnectedness of the business owner and the product owner and the designer and the developer and the tester, all are imperative to maintaining a constant pace. This is processual leadership in action. "One of the things that I think agile allows—and it's not perfect—it allows us to work with our business partners to understand the value of each requirement. And the business partner gets to hear about the complexity of what we're trying to build. Some requirements are more complex, yet they have smaller value, and so we can tweak the design. We can tweak the requirements to bring more value, bring things to market faster, reduce effort and cost" (Interview with Jeff Mander).

Credibility

Stakeholders and business owners need you to set realistic expectations because they are setting expectations of their own for *their* stakeholders based on what they can expect from you. Set realistic expectations both for your team and for your stakeholder. Your credibility is on the line. Better to risk achieving on expectations that to damage your credibility. Reputations are much more difficult to recover from than an occasional missed deliverable or missed expectation.

Business owners are less likely to push for "one more thing" and disrupt the process when you've established credibility, and they have the confidence that you will deliver with the next sprint, or maybe the sprint after that. When you've established credibility, your business owners can trust you and can count on you to do what you say you will do. It's simple, really. Dr. Ron Howard, Professor Emeritus at Stanford University, said it well: "Say what you mean, and do what you say."

Leadership in the Analysis Stage

You as the leader articulate the objectives for this stage: get clear on system requirements and get clear on where to start. In this stage, you ask specific questions. The objective of your questions is to help your team determine what's required of the system to deliver on the requirements that the product owner and the customer have spelled out. As important as determining what is required and an initial approach for achieving it, you and the team also need to determine alternatives and options. What other ways might the requirements be met? What options do you have? What alternatives exist? Is there prior work that you can leverage? Can you get a head start by finding similar solutions that exist elsewhere in your portfolio?

Leadership in the Design Stage

You as the leader understand and reiterate how important it is to frame the problem well. Framing the problem means getting clear on what you are trying to solve for and getting just as clear on what you are *not* trying to solve for. Effective framing reduces ambiguity and uncertainty. Any developer will tell you that they can't design and develop to ambiguous and unclear requirements. Without clarity, your developers are left guessing. And a product that is informed by best guesses is far less likely to satisfy the business owner and satisfy the customer than a product that is developed to clear requirements and clear expectations. At the Design stage, your team is determining *how* they will solve for the challenge from a technical perspective: what are the necessary interfaces, what will be required of the system, what will be required of the network, what are the data sources, what are the data flows? It's critical here that you as the leader ensure that the team documents its assumptions and why those assumptions were made. What informed the assumptions? Was it time? Was it system constraints? Was it scale expectations? Was it uptime requirements? The documentation is necessary for grounding the team in how and why key design decisions were made. Of course, documented assumptions also aid in traceability and in educating other teams down the road on why your team developed what they developed.

Leadership in the Development Stage

You as the leader keep your team true to the requirements that have been defined by the product owner, and you keep your team operating within established guidelines and constraints. Do keep in mind Principle 2, "Welcome changing requirements, even late in development." Recognize that expectations and needs will change. Your role as the leader is to ensure that your team understands this and works within the constraints of the system to deliver on these changing requirements. Developing software that no one will use does not serve your customer, it does not serve your team, and it does not serve you as the leader. Refer to Principle 2 for how best to do this in a manner that keeps you and your customer working together rather than at odds, and having constructive dialogue, not arguments, that generate useful and valuable outcomes.

Leadership in the Testing Stage

You as the leader ensure that your testing team knows what to test and why to test it. Ensure that the team has agreed with the product owner on the definition of "defects" and their severity. Remember, the objective isn't perfect software, the objective is valuable software that works. Perfect software doesn't exist. You as the leader are responsible for ensuring that your customer understands this and accepts this. If all your customer does when you explain to her that perfect software is a pipe dream, if all she does is nod her head, she may be simply accommodating you. Seek to understand her and seek to ensure that she understands you. Your customer doesn't need to know the technical intricacies of software development, but she needs an understanding of why perfect software isn't achievable. This understanding paves the way for user acceptance testing that is pragmatic and realistic, which in turn leads to the delivery of software that meets expectations.

Leadership in the Delivery Stage

You as the leader deliver code in iterations. This is the code that your team and the business owner have jointly planned. Your team and the business owner have jointly analyzed the business requirements and developed these higher-level business requirements, or user stories, into technical requirements. Your team has taken these technical requirements and performed the magic of software development. Your team has integrated testing into the development cycle and has worked closely with your customer throughout technical testing and user testing. You are ready to deliver the working software.

Leadership in the Maintenance Stage

You as the leader define for the product owner and for your customer what happens now. Software has been developed, tested, and delivered. Now it is operating, and without fail, your stakeholders and even your own developers will identify changes that need to be made. Your role is to ensure your team listens and responds. Seek input, acknowledge the input, and create a disposition for that input: will you implement changes? Will you add the desired changes to the backlog? Will you determine that the changes are significant enough to require entire development cycles and a major code release? Whatever your team's assessment, communicate this to your customer. If you can negotiate with your business owner exactly what needs to be changed, the magnitude of the changes, and the timing of the changes, this will help keep you aligned and working productively with your business owner. You may need to go to senior leaders and explain your decisions. Focus on what can be done and how, rather than on what cannot be done and why not. Remain solution oriented. Then set the right expectation for your senior leadership, for your team, and most importantly for your customer.

Key Takeaways

1. Listen.
2. Frame the work early and clearly.
3. You and your team must be healthy—physically, mentally, emotionally, and spiritually—to deliver at a consistent, predictable, and sustainable pace.

References

Carroll, L. (1893). *Alice's adventures in Wonderland*. T. Y. Crowell & co.
Gruver, G., Young, M., & Fulghum, P. (2012). *A practical approach to large-scale agile development: How HP transformed LaserJet FutureSmart firmware*. Addison-Wesley Professional.

Agile Principle 9: "Continuous Attention to Technical Excellence and Good Design Enhances Agility"

Abstract

Delivering valuable software frequently is the name of the game. Attending to technical excellence and to design quality can enhance agility. Create and apply automation selectively and strategically. Acknowledge and address defects and bugs and chronic performance issues in the software.

Vignettes from leaders at Oracle, Amazon, and UScellular illustrate the challenge of applying the discipline of technical excellence and good design to enhance agility.

Channel your team's technical prowess toward technical excellence and good design in service of delivering valuable software. Defects and chronic, customer-impacting issues must be fixed. Technical debt must be addressed. The objective of technical excellence, and the objective of good design, is the agility that enables you to deliver valuable software frequently.

What Does This Principle Mean?

What does this Principle mean? Here's a way to think about this: "We value excellence, and we strive for excellence. The framing matters. We insist on high standards. Technical excellence and good design are not the objective. The objective of this Principle is agility. We make practical use of technical excellence and good design. They are means to an end. And the end, the objective, is design and development that enhances the team's agility in delivering value."

Let's deconstruct this Principle.

"Continuous attention." Defining this broadly, this means more than "awareness." It means more than "consideration of." This means "persisting" and "unrelenting." It's a diligent and unremitting thoughtfulness to the technical excellence and good design that will enhance agility.

"Technical excellence." You as the leader must demonstrate that you know and can practice the distinction between "excellence" and "perfection," and between "good enough" and "best." Striving for excellence in technical skills and technical practices will improve the chances that the design will be excellent and good enough.

"Good design." What is "good" design? By definition, it is *not* "great" design. And "great" is not the impossible "perfect." "Good" in software development is code that is good enough. Design that is good enough allows work to continue without slowing down progress toward value. "Great" design might take too long and might distract the software development team from the mission of continuously creating and delivering working software. "Good" in software development is code that will do the job.

"Enhance agility." To enhance agility, create solutions that enable or even cause progress. Such solutions will be clear enough and simple enough. Technical excellence in the design means you will not be slowed down in your development because you have to fix fundamental or foundation problems.

You Are the Leader. What Do You Do?

Start with the End in Mind

What will this look like when you've achieved it? When this Principle is in place, your team will be creating designs and developing code that are grounded in technical excellence. They will not take short cuts that compromise future extensibility. They will design for the shortest paths for delivering working software that is valuable.

Listen and Learn from Others

How do you get to the "end in mind"? You start by asking great questions, and then you listen.

1. What is "technical excellence"?
2. How do you measure "technical excellence"?
3. What is "good design"?
4. What is "agility"?
5. How do you define "agility"?
6. How do you measure "agility"?
7. How can I help?

You've started by deconstructing this Principle to get at its meaning. You have defined the words and you understand the words. You've defined your objective with this Principle by starting with the end in mind. You've asked great questions and listened, really *listened*, to what you were told. Now it's time to learn from others.

When you think about design agility, and you think about how and what it means to apply continuous attention to technical excellence and good design, what specifically to you do? You are the leader. How do you approach these?

To set your team up for success in establishing and maintaining design, development, and delivery agility, apply continuous attention to:

- Technical excellence, with a focus on design reviews and automation
- Good design, with a focus on minimizing technical debt
- Enhancing agility, with a focus on design and development extensibility

Technical Excellence
Design Reviews
You as the leader can discuss with your development team the challenge of defects and the time required to resolve them, not as time devoted to fixing problems, but as time devoted to improving the customer experience. A defect by any other name would annoy a customer or impede the performance of the product or the code. But defects don't have to be viewed negatively, nor does the time required to resolve the defects need to be viewed negatively. You as the leader can position defects as the jumping-off point for doing even more for the customer. You as the leader can share your point of view that time spent resolving defects is time spent accelerating the customer's adoption of and enthusiasm for the product. Defects are not indictments of your team or of any one coder; defects are an opportunity to connect with the customer and commit to an improved product or experience, and ultimately to greater value.

You are the leader. Do you direct your team to carve out time to sit back and reflect and assess where they are and how they're doing, or do you empower and trust your team to do this without needing your explicit direction or permission? Consider Alex King's advice: "You as the leader must provide time and space to develop operational excellence, even if the 'customer' may not immediately see the results of that."

Maybe not immediately, but certainly sooner rather than later. Heather Ackenhusen shared the philosophy and practice of design reviews at Amazon. Consider:

"Design reviews should be part of an operations review process. So normally there would be a monthly operational review of the health of the infrastructure. We called these reviews Bug Bashes. These were days where we would get everybody together. We'll buy pizza. And everybody will do Bug Bash Day.

"Bug Bash Day could involve fixing a defect or a chronic issue or addressing what we would call 'paper cuts,' the thousand little issues were dragging the team down. These might be annoyances to the customer. These might build to become the difference between a product that customers seek out, and one that customers seek to avoid. Bug Bash was a way to build team back up by fixing things together. We're all in it together."

Good Design

Automation is one of many aids you can apply to help achieve technical excellence and good design.

Automation

Automation is an investment. You must measure and track the automation, like you would with any other investment. The testing team at UScellular creates value with automation in several ways:

1. Time savings—Using preproduction environments enables a team at UScellular to enhance testing by identifying possibilities for automation. They identify, develop, and apply those automations in the preproduction environment, where they next test them and optimize them. The team then applies those automations to production environments. The time saved through automation creates time for the team to address and resolve lower priority defects.
2. Process improvements—With fewer preproduction defects, the team has time to evaluate their testing processes. Based on what they learn, they either increase the depth of testing, or they transfer that testing to their regression suite. For production defect reductions, the team spends time creating meaningful analysis to share with the development teams to help them improve their standards, best practices, and designs for future projects
3. Capacity savings—The testing team accounts for the time savings that they realize with automation, which in turn enables them to absorb development or delivery delays, and to test more projects, for every code release.
4. Customer experience—The testing team continuously seeks ways to quantify the impact of defects on the customer experience. Today they are developing an automated process to map this impact. This mapping will enable the team to derive the value of each defect, which in turn will help them prioritize the work to resolve each defect. Today their process is cumbersome, but through automation the team is creating the labor capacity to design and develop an automated process.

Technical Debt

Teams begin to amass technical debt when their first solutions, arrived at through speed and brute force, are big successes right out of the gate. They work well and they work immediately. Good news, right? Not so fast. It's good if the solution was designed with flexibility and extensibility and scalability in mind. If the design accounted for all these, then the likelihood of technical debt decreases. But if the solution didn't account for these, not because the developers thought these didn't matter or weren't important, but because the fastest solution worked right away, then the risk of technical debt increases.

Technical debt can be insidious, creeping up on the team. Or it may be glaring and painfully obvious, but simply not as important to the team and to the customer as getting working software delivered. In either case and regardless of its origins, the rise of technical debt becomes a drag on the team and on the product. It becomes overhead. And that technical debt can lead to operational issues that distract developers, requiring them to divert their time and energy and attention away from development to troubleshooting. When they're troubleshooting, they're not developing. When they're not developing, they're not doing what they're best at. And when they're distracted from doing what they're best at, they lose motivation and you as the leader lose value.

Technical debt grows until it doesn't, meaning that it will only increase until the team diverts its attention from development to intentionally reducing technical debt. The opportunity cost of technical debt is significant. Technical debt can bog down agility. It becomes an anchor that the team must drag through the water.

Technical debt can happen when you're not paying close attention. Heather experienced this challenge.

Consider:

"You're moving quickly and you're experimenting. Sometimes you're not designing for the long term. You're doing a quick and easy implementation. Then suddenly you get caught because the product got super successful super quickly, and you discover that your implementation doesn't scale. So, then you're building a lot of manual stuff to deal with it.

"If you don't address that technical debt in a timely way, then your team could spend months trying to get back on a path of adding value or improving the customer experience.

"Be careful. Technical debt is something that you must care for."

Enhancing Agility
Extensibility

Continuous attention to technical excellence does more than provide a sound foundation for current projects. It also provides a foundation that is extensible for future work. Consider the case of the Linux Foundation. They are currently doing work they describe as Transition Analysis, which involves helping transition businesses to carbon-free operations. The engineers and software developers at the Linux Foundation are applying the discipline of data management to help solve the challenge of sustainability. The work involves treating the massive amounts of data that businesses generate as code. They do this by combining software development practices with data management practices to establish version control of the data and to manage the data lineage. This enables modeling and scenario planning from established and managed data sets. With help from the Linux Foundation, business across the spectrum and around the globe can see the impact of decisions they make today—and choices they make and the choices they don't make—on the sustainability of the environment.

Are You Ready?

You started by deconstructing this Principle to get at its meaning. You have defined the words, and you understand the words. You've defined your objective with this Principle by starting with the end in mind. You've defined what matters most. You know why this matters most because you've asked great questions of your team, your boss, and your customers. You've asked great questions and listened, really *listened*, to what you were told, and you have learned from their stories.

Now that you are clear on how continuous attention to technical excellence and good design enhances agility, what do you do? You are the leader. It is time to act.

Do

What does this look like in practice?

Continuous Attention
You seek feedback from your customer, and you act on that feedback. You seek feedback from your team, and you act on that feedback. You as the leader know where the risks are—chronic defects, technical debt—and it is your responsibility to create the time, the space, and the expectation to assess and address those risks.

Technical Excellence
If you strive for perfection, or if you demonstrate to your team that nothing short of perfection is acceptable, you will slow your team down, and you will frustrate them. Every good software developer knows that there is no such thing as perfect code. Code is developed to achieve, accomplish, or enable something specific. It is not developed to achieve and accomplish and enable everything. There is no such code and never will be.

Good Design
Heed the advice of the French philosopher Voltaire, who is credited with saying, "Don't let the perfect be the enemy of the good." If you declare that you have developed the perfect software or the perfect code, others will seek to find what is imperfect about it. On the other hand, if you declare that you have developed the code or the software that is good enough, your customer is more likely to accept it and work with it and understand it to be a work in progress that satisfies today's requirements and that can be enhanced and continuously improved *as necessary*. "Perfect" will cause your customer to be skeptical. "Good" will mean to your customer that the software works and is doing what they need it to do.

Enhance Agility
Adopt those approaches and those practices that enable you to adjust and move fast. Remember, though, that speed is not the name of the game. Delivering valuable software frequently is the name of the game.

Conceptual and Literal Thinking

You as the leader will be challenged by seemingly opposing forces of spending time, energy, and attention on technical excellence and good design, and on demonstrating speed and agility. I encourage you to adopt and practice two types of thinking: conceptual and literal. Learn when and how to apply them.

Conceptual Thinking

Your team needs to ask and answer conceptual questions. These include, "What does it mean to the architecture to add an important new module for the customer?" "What does it mean to the code base to introduce new features?" "With this new functionality, will we be creating technical debt?"

Your team as the technical experts need to engage with each other, with other coders and developers, other teams even, to ensure they are asking the relevant questions, and to ensure they are answering these questions incisively and accurately. You as the leader need to ensure this happens. You as the leader need to ensure that this generative dialogue takes place. It cannot become a distraction, and it must not impede progress. This dialogue needs to be processual—part of the process—and it needs to be generative—that is, it needs to lead to answers that determine next steps. The process of the dialogue must lead to decisions to do something or decisions not to do something. The decisions need to be explicit, and you as the leader need to ensure this happens.

Literal Thinking

Your team will be spending time and effort on the nuts and bolts of the design and the code. Your team has answered the conceptual questions— "What does this mean in terms of design and technical debt?" You must now ask and answer the questions specific to what the new modules and the new features and the new functionality mean to the code. "Can the code be extended?" "Can the code be refactored and be containerized for greater portability and extensibility, not only for these current design requests, but also for what might come next?" "Can the code and the underlying design scale as demand scales?"

You are the leader. As with the dialogue that you caused to answer the conceptual questions, you must now cause a similar rigorous dialogue to ensure that design considerations are part of the process, and that the process of dialogue generates new solutions or confirms prior solutions. Choosing not to decide is a choice. The point is the decisions need to be deliberate and explicit. They need to be *intentional*. Intentional decisions aid in traceability both for what was decided and for how the decisions were made.

Key Takeaways

1. Listen.
2. Technical excellence is not the objective. It is a means to an end. Delivering valuable software frequently is the objective.

3. Good design is not the objective. It is a means to an end. Delivering valuable software frequently is the objective.
4. Automation is not the objective. It is a means to an end. Delivering valuable software frequently is the objective.
5. Create and apply automation to assist you in maintaining technical excellence, good design, and delivery agility.
6. Create a discipline for seeking out bugs and defects and chronic issues. Assess the severity of each. If your customer tells you that the bug and defect and chronic issue are important, commit the resources to getting them fixed.
7. Agility is necessary for accommodating changing requirements and unexpected events. Requirements change and things happen, and you as the leader remain accountable for delivering valuable software frequently.
8. Conceptual and literal thinking will serve you in addressing the challenges with this Principle.

17

Agile Principle 10: "Simplicity—The Art of Maximizing the Amount of Work Not Done—Is Essential"

Abstract

Delivering valuable software frequently is the name of the game. The notion of simplicity needs to be applied to requirements, documentation, and priorities but not to code reviews. Simplicity reduces waste from the system and sharpens the team's focus on delivering what matters most and nothing more.

Vignettes from leaders at Amazon and Nokia, and a reminder from Albert Einstein, illustrate the challenge and the importance of achieving simplicity.

What Does This Principle Mean?

What does this Principle mean? Here's a way to think about this: "Keep it simple. As simple as possible but no simpler. The purpose is to eliminate waste, not only from the finished product, but also from the process. Waste of time, waste of attention, and waste of energy all are a waste of resources—all these are to be avoided and prevented."

Let's deconstruct this Principle.

"Simplicity." Simplicity is what remains when you strip away everything except that which matters most. When you set your sights on achieving maximum simplicity, you are both intentional and incisive. Intentional in leaving only what must remain and removing everything else. Incisive in the rigor and discipline you apply to determining what matters most.

"Art." Art is an expression or creation or application of skill and imagination. Art is not science. It is not formula-driven or algorithmic. It is achieved through thoughtfulness, imagination, humility, confidence, and experience.

"Maximizing the amount of work not done." Stephen Denning, in *The Age of Agile*, says it well: "Agile management is about working smarter rather than harder.

It's not about doing *more work* in less time: It's about generating *more value* from *less* work" (Denning, 2018).

"Essential." Essential describes what matters most. And what matters most to your customer becomes what matters most to you.

You Are the Leader. What Do You Do?

Start with the End in Mind

What will this look like when you've achieved it? When this Principle is in place, your team will be doing only what matters most. What *matters* most. Not "what's most fun," not "what's coolest," not "what's the most interesting or the most elegant of the most technically challenging." When this principle is in place your team will be doing what matters most to your business owner and your customer.

Listen and Learn from Others

How do you get to the "end in mind"? You start by asking great questions, and then you listen.

1. What does "simplicity" mean to you?
2. How do you achieve simplicity?
3. What do I need to simplify?
   (a) Requirements?
   (b) Documentation?
   (c) Software development?
4. What matters most to the customer?
5. How can I help?

You've started by deconstructing this Principle to get at its meaning. You have defined the words and you understand the words. You've defined your objective with this Principle by starting with the end in mind. You've asked great questions and listened, really *listened*, to what you were told. Now it's time to learn from others.

Let's start with Hamdy Farid's experience.

Consider:

"Agile doesn't have a monopoly on simplicity. When I went to engineering school, I learned how to write algorithms. The most beautiful algorithms—and I'm using the key word 'beautiful' because algorithms *are* beautiful when they're designed in the right way—the most beautiful ones are the simple ones. You want your solution to be simple."

How do you simplify in practice? You do three things:

1. Simplify requirements

2. Simplify software development, prioritizing what matters most
3. Simplify documentation

Simplify Requirements

Get clear on what needs to be done. Alex King, a senior engineer at Amazon Web Services, cautioned, "Watch out for a lot of work going into something before a lot of thought does."

You are the leader. Three things you need to do when you focus on simplifying requirements: stay connected, seek to understand, and stay focused on what matters most.

Stay connected with your team and your stakeholders. Why? Because needs and wants and necessities and requirements can evolve, subtly or significantly. You will only understand this if you stay connected with your development teams, your business owner, and your customer.

Understand what matters most. Think incisively. Be very clear on the value of each action. You as the leader must not tolerate any work that is extraneous. Don't encourage it, don't allow it, don't accommodate it, don't indulge it. There isn't time and space enough to do it, and there isn't value in doing it. Had you but world enough and time, but you don't. Understand your team well enough to know what matters most to them. Understand your business owner well enough to understand what matters most to her. Understand your customer well enough to know what matters most to them. Understand what is needed, and then prioritize what is needed most.

Is there a paradox here when you consider simplicity in the context of continuous attention to technical excellence and good design? I don't believe so. I believe that establishing simplicity as a guiding principle heightens your team's focus on what is truly excellent in their design. It leaves no room for the extraneous or the unintentional.

Stay focused on what matters most. How? Establish boundaries. Draw a line between what matters most and what doesn't matter most. What is most important to your business owner and to your customer is what will fall under the heading "What Matters Most." What goes under the heading "What Matters, But Doesn't Matter Most?" Everything else. Consider Henry David Thoreau's call to action: "Simplify! Simplify! Simplify!" Even this can be simplified. Fellow Transcendentalist Ralph Waldo Emerson responded to Thoreau's exhortation by saying, "One 'simplify' would have sufficed." Be more devoted to valuable outcomes than to anything else.

Simplify Software Development

Again, Hamdy Farid:

Consider:

"I think of simplicity in terms of agile as the process surrounding software development. The customer documentation could be simplified. I *wouldn't* simplify the code reviews. Code reviews really help the software engineers in reducing the number of code defects that go through the development cycle. I would rather have more code reviews up front to reduce number of bugs later."

Simplify development, but don't simplify code reviews.

Simplify Documentation
Hamdy continues:

Consider:

"Here's an example. Look at simplicity in customer documentation. When it comes to customer documentation and code documentation, the agile software developer can be very, very strict with it. He can write documentation for every line of source code. And then for every line of source code, you could write a comment. And for every function, you write a chapter in the customer documentation. But we know very well that much documentation is unlikely to be read."

Simplicity in documentation and in software code reviews. In documentation, aim for simplicity. In code reviews, *don't* aim for simplicity.

In the end, Hamdy says, "agile is finding the most amount of work that you don't have to do. Which is very clever."

Invest in Deeply Understanding the Problem
Many tradeoffs become much clearer when details are better understood. This is done several ways. Share. Demonstrate. Explain. Seek feedback. *Listen*. Ask the customer or the business owner *why* a particular feature is required, and *why* it is important that the user be able to see a particular field or perform a particular task.

Asking *why* is seeking to understand. Seek to understand the reasons for the requirements. Seek to understand the reasons for each desired outcome. Seek to understand the role of each user story. And *listen*.

The business owner will insist on getting everything *now* because they're afraid if they don't get it now, they may never get it. How often have you heard a product owner say something like this: "This is a great feature. Customers will really like this. This would be really great. A customer might want to x. I might want to x. Why *not* make this or that happen? It would be good to have in case we need it." The product owner asks for it because they might want it in the future. So why don't they just wait and ask for it in the future? Why ask now? Why muddy the water? Why risk slowing things down? Here's why: the business owner is afraid that the cool new future feature may not get funded in future rounds of funding or may be deprioritized in future rounds of development. This is an example of scarcity thinking. This is what product owners do to avoid scarcity thinking: ask for everything, hope to get most of it, settle of some of it. The mindset of the developer is this: tell me what you need and why you need it, and I'll get you something that works.

One indication that you truly understand something is that you can turn around and explain it or teach it. The same goes for requirements: If you and your team truly understand them, you can development software solutions to address them.

One key distinction that you as the leader must insist on and must continue to draw out. It's this: the distinction between "what it can do" and "how it will be used." This is the distinction between what's possible and what's necessary.

Another key distinction that you as the leader are responsible for understanding: the distinction between "need" and "want." What a business owner wants in their product, and what they need in the product, are likely two different things. What a customer says they need and what the customer really wants, again these may be two different things. You as the leader must guard against requirements sessions that become waking dreams.

Are You Ready?

You started by deconstructing this Principle to get at its meaning. You have defined the words and you understand the words. You've defined your objective with this Principle by starting with the end in mind. You've defined what matters most. You know why this matters most because you've asked great questions of your team, your boss, and your customers. You've asked great questions and listened, really *listened*, to what you were told, and you have learned from their stories.

Now that you are clear on what simplicity means and why it is essential, what do you do? You are the leader. It is time to act.

Do

You as the leader are charged with multiple and sometimes seemingly contradictory accountabilities. When facing trade-offs, ask detailed questions. Make it great, but get it done. Include us in your thinking, bring us along, but get it done, and now, please.

We all know that perfection can be the enemy of the good. You as the leader are charged with a particularly vexing challenge: how do you lead a team of smart, proud, highly skilled developers to do great work that is technically excellent, at the same time delivering *something that works* on schedule every time? Doesn't the perfect become the enemy of the good? One of the challenges that skilled and proud developers face is finding the balance between good enough and technically excellent.

Actively seek to avoid unnecessary work. Do better than simply eliminating it. Eliminating waste implies that it exists. Eliminating waste implies you're already let it build. Actively work to prevent it from existing. You as the leader must help the team prevent it. Remember, the goal isn't perfection. The goal is to create value for your customer. You do this by delivering working software frequently.

You can only know what matters most if you and your development team have engaged with your business owner and your customer early, and if you stay engaged with them daily. Don't guess and don't assume. Know what matters most, and then intentionally, deliberately, and purposefully do what matters most. Simplicity: this works and is necessary, whether you are leading a small team collocated in a small office, or you're leading a large, distributed, multinational team whose operation and whose work follow the sun.

The work of simplifying is captured well in this quote from Albert Einstein: "It can scarcely be denied that the supreme goal of all theory is to make the irreducible basic elements as simple and as few as possible without having to surrender the adequate representation of a single datum of experience." A precisely articulated summary statement from a 1933 lecture, simplified to "Everything should be made as simple as possible, but not simpler."

Key Takeaways

1. Listen.
2. What matters most to your customer becomes what matters most to you.
3. "Essential" is in the eyes of the customer.
4. Achieving simplicity requires you to be intentional and incisive.

Reference

Denning, S. (2018). *The age of agile: How smart companies are transforming the way work gets done*. American Management Association.

Agile Principle 11: "The Best Architectures, Requirements, and Designs Emerge from Self-Organizing Teams"

Abstract

Delivering valuable software frequently is the name of the game. The best architecture is the architecture that respects and accommodates customer requirements. The best requirements are simple and clear. The best designs enable the customer to do what they need to do the way they want to do it. Taken together, the best architectures, the best requirements, and the best designs comprise a solution that enables the customer to use the product the way they want to use the product. The best solutions are sensible, scalable, hardy, and timely.

Vignettes from leaders at UScellular, TDS Telecom, Nokia, Amazon, and Oracle illustrate how the best architectures, requirements, and designs can emerge from self-organizing teams.

What Does This Principle Mean?

What does this Principle mean? Here's a way to think about this: "We've got this. We self-organize, and we're best this way. Our team *knows*. Get the right people on our team, and we will *know* the architecture that's best for the product. Get the right people on our team, and we will *know* how to structure requirements that will propel us and not distract us. Get us the right people on our team, and we will *know* how to develop the best designs." The key notion common across each of these actions is "emerge." Get the best people and trust them to self-organize, and bests—best architectures, best requirements, best designs—will *emerge*.

Let's deconstruct this Principle.

"Best." For our purposes, this means "best suited," *not* "best ever."

"Best architectures." From Gruver, Young, and Fulghum: "The best architectures are those that include structured interfaces, extensible designs, callable microservices, and plug-and-play modular designs that maximize the efficiencies and the agility that you get with containers. These characteristics help your development team establish a core to work from, a source, a foundation for tomorrow's work. These characteristics also help you maximize your business owner's investment" (Gruver et al., 2012).

"Best requirements." The best requirements are those that emerge through dialogue. The best requirements are those that address your customer's needs. Best requirements are clear, easily understood, and expressed in such a way to cause progress toward a product that delights your customer.

"Best designs." Back to Gruver, Young, and Fulghum: "The less rework and the more reuse, the less the throw-away investment. Optimize the design so both you and your business team can leverage existing solutions to help maximize the investment" (Gruver et al., 2012).

"Emerge." This is what happens when you join people together in a process that generates a valuable outcome. "Interacting agents contribute to an ongoing process of 'collaborative emergence' (Sawyer, 2005)" (Tsoukas, 2017).

"Self-organizing teams." These are teams who organize their skills and capabilities to solve the problem at hand. Teams know themselves, so teams know best how to organize for optimal performance.

You Are the Leader. What Do You Do?

Start with the End in Mind

What will this look like when you've achieved it? When this Principle is in place, your team will be meeting with the people they need to meet with, talking with the people they need to talk with, discussing and resolving the points that require discussion and resolution, and they'll be doing this without requiring your intervention. They will have translated a requirement to a user story and a user story to an outcome. They will architect and they will design with that specific outcome in mind. They will iterate. They'll be solving problems and making progress, and they'll keep you informed.

Listen and Learn from Others

How do you get to the "end in mind"? You start by asking great questions, and then you listen.

1. What makes an architecture "best"?
2. What makes requirements "best"?
3. What makes a design "best"?
4. Who gets to decide the answers to 1–3 above?
5. How can I help?

You've started by deconstructing this Principle to get at its meaning. You have defined the words and you understand the words. You've defined your objective with this Principle by starting with the end in mind. You've asked great questions and listened, really *listened*, to what you were told. Now it's time to learn from others.

The Best Architectures

The best architectures are the result of clear understanding of the problems to be solved today and an informed perspective of what might need to be solved tomorrow. The best architectures emerge from rigorous thinking and rigorous debate. They are clear. They make clear the direction that solutions will take. The best architectures are not so rigid that they preclude creativity in the designs of new solutions. They guide. They don't dictate or predetermine.

The Best Requirements

Clint Wallin is a Senior Director at UScellular. In his role leading the development and delivery of software and platform solutions, he has significant experience with requirements definition. "A 'best requirement' would be one that the team collectively understands. The team understands what needs to get done and why." This is a critical distinction: the best requirement is not so high-level that it's nebulous, and it's not so basic that it expresses only a single command in isolation. The best requirements in Clint's view are *practical* and are easily *understood*.

Consider:

"A depth of understanding is achieved and realized by seasoned teams, by teams that have been together for months or years. Why? They've learned to communicate effectively, and they've established deep trust. Experience and trust within a team enables high performance. The team won't need a lot of detail in the requirements. I think they actually need very little because of the dynamic of the team. The people who write requirements understand how their business partners think because they know them so well. It becomes seamless. And if they miss something and there are gaps, they know enough to fill them in.

"They know each other. They've spent a lot of time together, so they know how to write requirements in a way that will be interpreted correctly.

"This is how it works for a high-performing team. If the team is new, or if it's heavily outsourced to third parties where the people on the development team change frequently, you'll need a lot more detail in the requirements in order for people to understand the context and the meaning implied by the requirements."

The Best Designs

Recall Mike Brendzal's comment in Principle 10 on this topic. The aim is not to do more work in less time, but to deliver more value with less work. Back once more to Gruver, Young, and Fulghum: "We found it extremely important to align code architecture with business objectives and architect the code so that, wherever possible, we could eliminate non-value-added work from the system. The approach to architecture and the directly related approach to coding standards can either lead you in the wrong direction or provide significant breakthroughs in development or deployment efficiencies" (Gruver et al., 2012).

Clint Wallin puts it succinctly: "The best design meets requirements. It is simple, reusable, and easily supportable. You could come up with a really cool, clever design that's so complicated that nobody can understand it, and nobody can support it, and you can't reuse it for anything else."

And from Karl Betz with TDS Telecom.

Consider:

"In agile, 'best designs' means we're designing things in a way to be flexible. We're designing them to be loosely coupled. We're designing them to be easily adaptable so that we're not hardwired into a certain predetermined design, but we have flexibility to adapt it as those user stories evolve over time and as the users themselves start to give you feedback. Then the technology people start to go say, 'I really understand where we're going with this and how we can adapt the design.' The best design is one that is flexible and adaptable."

Who will do this aligning and this architecting? Who will do the elimination of non-value-added work? Who will code? Who will achieve significant breakthroughs? The self-organizing team.

The Best Self-Organizing Team

Self-organizing does not mean self-building. You as the leader are responsible for building the team. You are responsible for the composition of the team, the skill sets, the experience, the thought diversity. Self-organizing is not the same as self-selecting. The team doesn't select itself or build itself. You as the leader are responsible for the composition of the team that is charged with delivering valuable software frequently.

Your role as the leader is to define the objective, get resources, minimize distractions, and establish priorities. Your role is to establish the conditions that enable your team to deliver. These are the conditions—support, clarity, resources—that enable your team to perform at the highest levels.

As you build the team, establish criteria for what makes the best software developer, the best product owner, and the best business owner. What do I mean by "best"? Not "best ever." Instead, "best suited." Best suited to your organization, its mission, the business requirements, and the customer expectations. There is a critical distinction between "best ever" and "best suited." You as the leader need to know the difference. Insist on the best. Insist on high standards. Then determine the context. Determine the application. Determine the stage of development. Then get to work building your team.

Again, Clint Wallin and his perspective on what makes the best self-organizing team:

Consider:

"The best self-organizing team is made up of people who are multi-skilled, humble, and outcome focused.

"You need to have people on the team who can do many things. They can wear many different hats. They don't have to be experts in each area, but they have to be able to do some activity in every aspect that's needed, whether that be requirements, design, development, or testing. They may or may not have all those skills, but they need to have more than one skill.

"And then humility. They don't let their pride get in the way of getting work done. Just because I may be a really good developer, an expert even, if some basic flow needs to get tested, I'm going to jump in and get it tested. If somebody needs to sit down and write a requirement, I'm going to sit down and write a requirement. Whatever the activity is, if it needs to get done, I'm going to jump in and take care of it, even though it may be something that I did 10 years ago in my career. I'm not going to let my pride get in the way. I'm going to take a humble approach to just getting the work done.

"And that's where that outcome focus then comes in. I'm less focused about a particular activity and I'm more focused on are we going to deliver on the outcome that we agreed to as a team. And we're going to keep our capacity and our throughput and everything else going the way that we need it to. So, when I think of a self-organizing team, those are the three attributes that I think of: multiple skills, humility, and a focus on the outcomes."

We've defined best architectures, best requirements, and best designs. We've defined what a self-organizing team and the characteristics of the best teams. But *who* is on this team? What roles comprise the team? Karl Betz shared his view.

Consider:

"The self-organizing team is made up of not only developers, but it's also made up of users and it's made up of product owners. It's made up of testing people and database people.

"The self-organizing team will listen to the end users. They're going to listen. And they're also going to carve out time to evolve the design and invest in the design. It comes back to the self-organizing teams recognizing the importance of an architectural design that provides the capabilities by which you can more quickly and easily deliver value to the business.

"If you have a truly self-organizing team, if they're truly autonomous and they're empowered, they will prioritize what's most valuable. The best user story and the best requirements will rise to the top. Self-organized teams will constantly prioritize and organize their work for the greater good of the business."

The Best Software Developers

What makes the best software developer? The best software developer, according to a leader at a world-class software development company:

Consider:

"I feel that the ideal software developer is a person who understands the fundamentals of software programming. This person has a good attitude and has a learning attitude. That means they like to explore, find good technologies, find what fits the solution. If they don't know something, they work hard to learn it. They pay attention to quality.

"I don't put too much emphasis on whether they know a certain technology or whether they have a done the work earlier. It's more important that they ask questions. They should be able to collaborate. They should be able to learn, explore, and build software consistent with the organizational environment."

Beyond having the skills, characteristics, and attributes that Clint and others call out, the best software developers need to understand the business. They need to understand the context for what they're being asked to develop. They need to understand the business owner's point of view. Consider what Hamdy expects of the best software developers:

Consider:

"I need the developer to understand what problem statement we're trying to solve. If my developer understands the problem that they're solving, everything will fall into place. If the developer doesn't understand and continues to create something that has limited or no value to the business owner or the user, that will create a conflict. If there is one thing that characterizes the best developers and that I need from those developers, I need them to understand, 'Why are we solving the problem statement and who is on the receiving end of the solution?'"

The Best Product Owners

I asked one leader how she would describe the ideal product owner. She laughed, "That's something I wonder myself." Then she explained: "I think that *ideally* a product owner is a domain expert combined with a strategic decision-maker combined with a customer expert. This person needs to be able to decide what needs to be done and what is important at a given time and how the market is moving so that they can take full ownership of delivering and have a stake in the success of what we are delivering. They should have authority and a good understanding of the domain as well as what we want to deliver to the market. She will understand her product. She will understand her customer. She will understand the value of clear requirements."

The Best Business Owners

We've identified the importance of software developers understanding the point of view of the business owner. This is one of the defining characteristics of the best developers. But what makes the best business owners? Hamdy Farid:

Consider:

"The thinking process and the basic development principles. A good business owner is someone who began as a developer. And in a technical industry like mine, it's useful for the business owner to have a technical background.

"I can tell you that the best business owners come from a technical background. They don't have to understand agile. They don't have to understand SAFe. They don't have to understand all the types of development. But they do have to appreciate it. The best business owners appreciate how the engineer's and developer's brain operates."

Are You Ready?

You started by deconstructing this Principle to get at its meaning. You have defined the words and you understand the words. You've defined your objective with this Principle by starting with the end in mind. You've defined what matters most. You know why this matters most because you've asked great questions of your team, your boss, and your customers. You've asked great questions and listened, really *listened*, to what you were told, and you have learned from their stories.

Now that you are clear on what it means for the best architectures, requirements, and designs to emerge from self-organizing teams, what do you do? You are the leader. It is time to act.

Do

Your role as the leader of the self-organizing team includes building the team, setting the direction, and establishing expectations for their performance. Your role also includes making it safe for the team to explore and making it safe for the team to make mistakes. Provide opportunities for your team to think outside the box and identify problems outside their immediate space. Make it safe for your team to test, to try, to explore, to consider. Make it safe for your team to remain curious about their domain and actively seek areas for improvement, even if they are not on the product roadmap.

Your team takes it from here. Your team is responsible for determining how best to assign tasks and how best to track their work and how best to report out on progress and how best to interpret their progress. Your team is responsible for raising its collective hand when it needs your help. Your team is responsible for holding each other accountable for daily progress. Your team is responsible for knowing the tools they need—your role as the leader is to procure them. Your team is also responsible for seeking and responding to feedback. Their role is not simply to do what you tell them. They must listen to and respond to the feedback from the business owner and from their customer.

Jeff Mander describes how this works at UScellular.

Consider:

"The scrum master holds everything together, making sure they're the ones leading that daily standup. They're the ones leading the sprint planning sessions. They're the ones holding the retrospective and then managing all the output. They're making sure that we're going through each story, working with the developer, making sure that the developers are presenting the story successfully according to what the business owner and the product owner wanted. The scrum master acknowledges the feedback and takes the feedback and makes sure that the team works it in.

"Your team decides how to best structure, how to best allocate work, how best to report, how best to adjust. They don't need you as the leader trying to control changes to any of this. They don't need you swooping in to save the day. They don't need you calling plays from the sideline. They need your direction, they need your support, they need you to remove obstacles, they need you to listen, and they need you to stay out of the way."

The Distinction Between "Self-Organizing" and "Self-Managing" Teams

What's the difference? Is there a difference? If there is a difference, does the difference matter, or is it just semantics? Jeff Mander will tell you that there is a difference and that the distinction matters.

Consider:

"I like the term 'self-managing' better than 'self-organizing.' The leader puts together the team, and then they self-manage. This means that the product owners are working with the business team, and the business team is giving them the priorities, giving them the user stories, giving them guidance on what they're looking for. Then the development team is asking and determining, 'What can I do? How much effort is it? How am I going to break down these tests to complete what I need to do to meet that the requirements of that story?'

"The self-managing team also self-manages their learnings. They take learnings from the retrospectives and assess the learnings and figure out what to apply and how to apply it. The scrum master is integral to that part, making sure that he or she is acting as a servant leader, removing any obstacles that the teams have, coordinating with the product owners as needed to resolve any open questions that the team might have as they're building. They're leading demo sessions and they're making sure that the team is listening to the feedback and applying that feedback into the development cycle.

"To me, that's what self-managing teams do. I'm the leader, so I build the team and organize it. Then the team does the hard work. That's why I like the term 'self-management' more than 'self-organizing.'"

Deciding When to Empower Self-Organization

When *does* it make sense to empower self-organizing teams with the choices and decisions on architecture, requirements, and design? It depends on the context.

Context

The choice to empower self-organizing teams with this work depends on context. In determining the context, the scale of the project that the team is going to work on is the first consideration. Scale is important. That's Jeff's first question in determining whether he can have self-organizing teams do the development. In his experience, self-organizing works well on smaller projects.

Can agile development teams, on their own, determine the best designs and the best architectures? Can they do this universally, or is it context-dependent? In Hamdy's experience, they can do it.

Consider:

"If you put together a group of architects and you give them a task of coming up with the architecture, you're trusting them to get the job done. They can do this. The initial design needs to be done by a smaller central team.

"As the business owner, I don't want the developers to come to me for design decisions. I want them to come to me for features and usability requirements, but not for the design of the product architecture. When I go for an architecture review, I ask a few simple questions: 'How does it scale? How does it fail if I throw more workload at it, what will happen?' I need to know the answers to those questions. I'm also curious to understand architecture because I am an engineer. But having input on the architecture? That's not what I ask about. The questions I ask are, 'Does it scale well to handle load? How does it fail? Does it fail gracefully?'

"These are usability questions. If all of them check out well, then great. Later, and offline, I can go to the lead architect and say, 'Can you please have a one-on-one with me? I would like to understand the design more deeply because I am curious.'"

This reminded me of a conversation I had with one of the senior engineers at Amazon. His view was similar to Hamdy's. He emphasized with me that what matters in the design is how well does the solution scale, how well does it handle load, does it have the capacity to handle radical increases in traffic without breaking?

Empowering people does not mean that leaders step out of the picture entirely. They still have a critical part to play that includes establishing boundaries. This generally takes the form of establishing rules, roles, goals, and measures for the organization. Constraints need to be established and communicated up front. Boundaries need to be closely monitored and loosely managed to allow groups to self-organize and develop solutions. Managers can then adjust as necessary to raise or lower the amount of energy interjected into the system to guide processes, problem solving, and innovation.

You are the leader. Your role is future oriented. What do you do relative to the "best architectures"? Your role is to ensure that the business owner makes the product roadmap clear. Your role also includes understanding and then sharing with your team the longer-term roadmap for the product. You as the leader will influence their work but not control it. You need to know and to *share with your team* what the plan is, and why the plan is what it is. How does this product fit into the long-term roadmap and into the strategy and into the vision? How does it fit, and why does the company believe it is worth its investment and worth *your* investment and your *team's* investment?

You as the leader must be able to articulate for your team the compelling "why" for doing this work. Technology teams are motivated in part by understanding the compelling "why" for the work they are doing. A compelling "why" goes a long way to keeping them motivated and engaged and investing discretionary effort in their work. (Refer to Principle 5 for a more fulsome discussion of why the value of the work matters to a software development team.) (Also refer to Principle 10 on simplicity.)

Your role as the leader is to trust your team. Trust and verify. Trust and inspect. Trust and support. Trust and deliver. Trust and celebrate.

Key Takeaways

1. Listen.
2. Your role as the leader is to find the talent, attract the talent, get the talent, and retain the talent that your team needs so that they can do their job.
3. Your role as the leader is not to determine the best architectures, requirements, or designs. These are your team's responsibilities.
4. Your role as the leader is to enable great outcomes.
5. Your team's role is to create and deliver great outcomes.

Reference

Gruver, G., Young, M., & Fulghum, P. (2012). *A practical approach to large-scale agile development: How HP transformed LaserJet FutureSmart firmware*. Addison-Wesley Professional.

Tsoukas, H. (2017). Don't simplify, complexify: From disjunctive to conjunctive theorizing in organization and management studies. *Journal of Management Studies*, 54(2), 132–153.

Agile Principle 12: "At Regular Intervals, the Team Reflects on How to Become More Effective, Then Tunes and Adjusts Its Behavior Accordingly"

Abstract

Delivering valuable software frequently is the name of the game. Teams can improve. They can be told what to do and how to do it. Or they can be trusted to engage in honest self-reflection. Honest self-reflection becomes the best and most effective means for the team finding and creating sustainable improvements in the way they operate.

Vignettes from leaders at TDS Telecom, Oracle, and Calstate Management Group, Inc., and a reminder from Richard Feynman, illustrate the challenge and the importance of regular reflection.

You as the leader must encourage but not require self-reflection among your team. Enable it without prescribing how they should do it. Self-reflection can lead to discovery, and discovery is the key to continuous improvement.

What Does This Principle Mean?

What does this mean? Here's a way to think about this: "We are curious and open-minded. We are honest with ourselves and each other. We know how we work. We know our strengths and our weaknesses. We are not anchored to ways of working. We are anchored to delivering value. We reflect, we adjust, and we move on. We make progress every day."

Let's deconstruct this Principle.

"At regular intervals." What are "regular intervals"? Agile software development teams are expected to move fast, minimize waste, and deliver continuously. All this, but then ask them—*expect* them—to slow down, pause, and take time to reflect. Reflection is essential. What are "regular intervals"? Let your team decide and then insist that they adhere to this.

"The team." This is everyone. This includes you as the leader. This includes the product owner. This includes the business owner. This includes your developers.

"Reflects." Think of this as an honest assessment of yourself and your team that in turn leads to a disciplined exploration of new solutions. John Young, Group President of Pfizer Essential Health, said in an address that "reflective thinking improves decision making by grounding it in a more integrated and coherent world view than one can have from acting only in the moment." This requires focus and attention and effort and discipline and commitment and energy.

"How to become more effective." We'll define this narrowly. "Effective" is frequently delivering working software that adds value.

"Tunes and adjusts its behaviors." Tuning and adjusting behaviors is a form of innovation. Innovation comes in all shapes and sizes. Reinvention is innovation. Creation is innovation. Adjustments are examples of innovation. They can be formal and applied change management initiatives. They can be subtle and nuanced changes in behaviors.

"Accordingly." According to what? According to how best to deliver value in a sustainable way. In a way that respects each team member. In a way that generates growth.

You Are the Leader. What Do You Do?

Start with the End in Mind

What will this look like when you've achieved it? When this Principle is in place, your team will be reflecting regularly and will be learning continuously.

Listen and Learn from Others

How do you get to the "end in mind"? You start by asking great questions, and then you listen.

- What does it mean to reflect?
- Why does reflection matter?
- How do you make reflection worthwhile?
- When you reflect, what do you reflect on?
- What do you believe is possible?
- What's it worth?
- How can I help?

You've started by deconstructing this Principle to get at its meaning. You have defined the words and you understand the words. You've defined your objective with this Principle by starting with the end in mind. You've asked great questions and listened, really *listened*, to what you were told. Now it's time to learn from others.

What Do Individuals Do?

They buy in to a self-critical mindset and always try to disprove their own ideas. Karl Betz shared his view on team reflections. His team reflects on what they do and how they do it. They look for opportunities for continuous improvement. They look critically at what works and why, and at what doesn't work and why. Between the two, they aim to identify opportunities for improvements in speed, agility, cost, cohesion, and quality. They look for opportunities to make qualitative improvements and quantitative improvements. For example, they may find that they can retire tools and environments. Are there DevOps tools that can be adopted? Can they retire legacy software repository solutions? Can they reduce the age of data across their environments?

What Do Leaders Do?

Muhannad Obeidat described the role of the leader relative to the team's self-reflection. "The leader has a role to play. The role of the leader is encouraging and driving innovation. You as the leader don't need to be in the meetings with the teams, but you should encourage the team, even expect them, to have these meetings."

Muhannad describes his practice. "I try to inspire people to think differently about how we do things. I try to inspire the team by bringing ideas from outside our immediate area of concern and then inspire the team with these ideas."

You as the leader can create the opportunity for formal reflection by doing what Murray Krehbiel does for his team. Murray is the CEO of Calstate Management Group, Inc. "Even though my team is not set up as an agile team, we embrace many agile principles including holding recurring retrospectives. We've drawn upon some of our formal agile scrum training to put this in practice at the end of each of our 2-month reporting cycles to call out opportunities for improvement. This is a good practice for any team even if the team is not a formal agile software development team."

Why reflect? Why ask your team and expect your team to spend its valuable time, energy, and attention in reflection? You do this because of the value it can generate. From Uhl-Bien and Arena:

Consider:

"Leadership for organizational adaptability focuses on enabling the adaptive process in organizations. At its core, the adaptive process is about engaging the tension between the need to innovate and the need to produce (March, 1991) …Perhaps the bigger challenge in engaging this tension is in not letting the pressure to produce overwhelm the need to innovate. Because most organizations are designed for stability, they are proficient at rejecting new ideas and change (Leonard-Barton, 1992). The formal structure is designed to suppress the informal structure of networked interactions. Moreover, managers are trained in hierarchical leadership with a bias toward order and a focus on top-down control" (Uhl-Bien & Arena, 2018).

You are the leader. As the leader, you owe it to your team to show you trust them. Give them time and space and permission to reflect. The output they generate in reflection becomes their opportunity to improve.

Gisela Backlander's research on the practice of agile coaches at Spotify emphasizes the opportunities that self-reflection can generate:

> Research on teams shows that team learning behaviours—e.g. 'sharing, discussing, and reflecting on knowledge and actions' (Koeslag-Kreunen, Van den Bossche, Hoven, Van der Klink, & Gijselaers, 2018) or 'asking questions, seeking feedback, experimenting, reflecting on results, and discussing errors or unexpected outcomes of actions' (Edmondson, 1999, p. 353), and not least team reflexivity (e.g. overtly reflecting on and communicating about goals, process, and outcomes; Schippers, Edmondson, & West, 2014)—are related to innovation (Schippers, West, & Dawson, 2015; Widmer, Schippers, & West 2009) and other adaptive outcomes (Mathieu et al., 2008). (Backlander, 2019)

Your team's self-reflection is their opportunity to transform in small ways or large, with the objective of improving its efficiency and effectiveness in delivering valuable software. "Outcomes are compared with ideals or plans and are fed back as inputs (plans) for the next iteration" (Tsoukas, 2017). Think about the agile software development process. Recursive operations—think 'sprints' in the world of agile software development—are opportunities for creativity. "The experience generated through the recursive performance of the routine potentially changes agents and, therefore, the way future iterations of the routine may be enacted" (Tsoukas, 2017).

Be mindful of limiting your field of outcomes. Adherence to existing patterns risks excluding the new, the innovative, and the different.

Are You Ready?

You started by deconstructing this Principle to get at its meaning. You have defined the words and you understand the words. You've defined your objective with this Principle by starting with the end in mind. You've defined what matters most. You know why this matters most because you've asked great questions of your team, your boss, and your customers. You've asked great questions and listened, really *listened*, to what you were told, and you have learned from their stories.

Now that you are clear on what it means for your team to reflect at regular intervals on how to become more effective, on how to reflect for effectiveness, and then tune and adjust its behavior accordingly, what do you do? You are the leader. It is time to act.

Do

You as the leader need to encourage your development teams to take time periodically to push away from their keyboards, stand up, stretch, take deep breaths, and reflect. You can make this a formal part of the process like Murray Krehbiel does. Or you can encourage it and create space for it, similar to Muhannad's practice. Regardless of your approach, what matters is that your team spends the time, energy, and attention in reflection.

Their reflection should include questioning why they do what they do the way they do it, the tools they use, the processes, the meetings, the stand-ups, the demos, the ceremonies, everything. You as the leader need to create the time and the space for them to do this. And the team needs to approach this intentionally and with a sense of healthy dissatisfaction.

I use the phrase "healthy dissatisfaction" deliberately. A healthy dissatisfaction is not the same as frustration. It is not the same as cynicism. It is not the same as discontent. Healthy dissatisfaction is a source for change and for innovation. It can be a source of creativity. It is the mindset that meets a challenge with optimism and with energy. It is the mindset that determines to make change to the current state with the confidence that the current state can become a better state. "There is no harm in doubt and skepticism, for it is through these that new discoveries are made" (Richard Feynman, Nobel Laureate). Healthy dissatisfaction envisions tomorrow, assesses today, and determines that the vision of tomorrow will become a reality. That's healthy dissatisfaction.

The most effective leaders must start by demonstrating a strong commitment to a new direction but admit that they do not have all the answers (Kotter, 2012). They should recognize the need for change and guide the organization towards it while not imposing specific actions. Effective leaders recognize that the people closest to problems often have good ideas for addressing them. They accept that implementing planned change in organizations is always an experiment in progress and that different approaches should be tried to find out which ones work best. Effective leaders identify opportunities, establish boundaries, encourage the engagement of system agents, and then allow the change to unfold (Kelly & Allison, 1999).

Any development team worth its salt will challenge itself to get better than it's ever been. It believes in itself. It believes it can be better, faster, more efficient, more effective. It believes it owes this improved state to each team member and to the broader organization. It operates with a sense of opportunity and a sense of obligation to the broader organization. If you as the leader have a team that does not believe that better is possible or that better isn't worth the trouble, then you have the wrong team and you need to make changes. On the flip side, if you have a team that believes it is unstoppable, that "better" isn't simply *possible* but is *necessary,* and it is *in their hands to make it happen*, then you've got an "A" team. Set them up, support them, celebrate them, and get out of their way.

In one sense, "positioning people to be adaptive" is no more than letting the team do their work and staying out of their way. Consistent with this agile Principle, enable, allow, and encourage the team to reflect, consider, discuss, and debate. Give them the time and the space and the encouragement to do this. You as the leader must find time in the pressures of the delivery schedules to allow for your team to reflect. A team of empowered professionals will act in accordance with the Principle: they will figure out how to be more effective, and they will change how they work. They will, as the Principle states, "tune and adjust." This is innovation. You as the leader don't *cause* the innovation; you *enable* it.

What does this time spent in reflection get you? Maybe nothing. Your team may identify an opportunity but determines that the gain isn't worth the pain and that the juice isn't worth the squeeze. Or they might generate stability through process improvement. They might generate innovations, asking and answering the question of what's possible, asking and answering the question of value. They will not know until they try.

Key Takeaways

1. Listen.
2. Think deeply.
3. Reflect honesty. Dishonest reflection is a waste of time. Honest reflection can help you imagine what "better" looks like and can help you find options for achieving it.

References

Backlander, G. (2019). Doing complexity leadership theory: How agile coaches at Spotify practice enabling leadership. *Create Innovation Management, 28*, 42–60.

Kelly, S., & Allison, M. A. (1999). *The complexity advantage: How the science of complexity can help your business achieve peak performance*. McGraw-Hill.

Kotter, J. P. (2012). *Leading change*. Harvard University Press.

Tsoukas, H. (2017). Don't simplify, complexify: From disjunctive to conjunctive theorizing in organization and management studies. *Journal of Management Studies*, 132–153.

Uhl-Bien, M., & Arena, M. (2018). Leadership for organizational adaptability: A theoretical synthesis and integrative framework. *Leadership Quarterly*. https://doi.org/10.1016/j.leaqua.2017.12.009

Part V

What's Next

Conclusion

20

Abstract

Leadership is a social construct. It is conjunction, processual, and generative. It is best practiced with a keen eye and a kind heart.

To lead modern technology teams, we need to think in terms of action. And the actions we need to think about are those actions that are conjunctive, processual, and generative. We need to train our leaders and ourselves in terms of conjunctive leadership, processual leadership, and generative leadership. Leadership is not an entity; it is a process. And the process of leadership needs to generate progress both in thought and in action. Leadership must enable and then cause something to happen.

On the Principles: A principle is not an objective. A principle is a guide, not a destination. It is not an endpoint but rather a jumping-off point.

We can always find reasons why things don't work. We can mock what someone else believed in. We can say, "I told you so." We can look at a principle and find reasons to deny it. If we look hard enough—sometimes, we don't have to look all that hard—we can find the fissures and the cracks. If we look hard enough, we can find failures. We can point out inaccuracies and imprecision. We can find what we consider inappropriate, too simplistic, misguided even. Said differently, if we look critically, we will find what we're looking for. Look for a fault, and you will find it. Especially when the statement is aspirational. There's nothing easier than to knock the idea that aims high, to shake our heads and chuckle at the naivety of the aspirational statement, or to dismiss the idealistic as simple-minded.

Robert Browning got it right when we wrote, "Ah, but a man's reach should exceed his grasp, or what's a heaven for?" I applaud the people who aim high. I applaud the principles and I cheer on the people who wrote them. I will stand on their shoulders. I will build on their foundation. I believe that we can improve, that we can be better for ourselves and for the people around us, especially for those people who depend on us and whom we can serve. I choose to believe in the power of principles. Instead of pointing out what doesn't work, can we seek to find what *does* work and how we might contribute our own verse, our own thinking, our own values and beliefs and make it even better? And looking forward, can we hope that one day, someone down the road, in a future that we cannot see, will look thoughtfully and critically and generously at what we wrote and what we said and what we believed, and find the beauty in it, and imagine and create ways of improving on it?

Let's look at the principles from twenty years beyond when they were written. Let's find what works and how it works. And let's see what we might challenge, not with the objective of dismissing it or discrediting it or throwing it away, but with the objective of seeing what works, what we can apply, what we might extend to make things a little bit better. That should be our aim. See critically but generously. Incisive but not derisive. Not destructive but generative. Seek not to destroy but to create, to build, to make a little better. And hope that someday, someone will approach our work the same way. See what's beautiful and see what's possible.

Bibliography

Arena, M., & Uhl-Bien, M. (2016). Complexity leadership theory: Shifting from human capital to social capital. *People and Strategy, 39,* 22–27.

Backlander, G. (2019). Doing complexity leadership theory: How agile coaches at Spotify practice enabling leadership. *Create Innovation Management, 28,* 42–60.

Bloom, N. (2020). *How working from home works out.* Stanford Institute for Economic Policy Research (SIEPR).

Bloom, N., Han, R., & Liang, J. (2023). How hybrid working from home works out. *National Bureau of Economic Research Working Paper Series.* https://doi.org/10.3386/w30292

Burger, A. (2018). *Witness: Lessons from Elie Wiesel's classroom.* Houghton Mifflin Harcourt.

Carroll, L. (1893). *Alice's adventures in Wonderland.* T. Y. Crowell & co..

Collinson, D. (2014). Dichotomies, dialectics and dilemmas: New directions for critical leadership studies. *Leadership, 10,* 36–55.

Coursera. (2022). Retrieved from Coursera: www.coursera.org

Denis, J.-L., Langley, A., & Sergi, V. (2012). Leadership in the plural. *The Academy of Management Annals, 6*(1), 211–283. https://doi.org/10.1080/19416520.2012.667612

Denning, S. (2018). *The age of agile: How smart companies are transforming the way work gets done.* American Management Association.

Dionysiou, D. D., & Tsoukas, H. (2013). Understanding the (re)creation of routines from within: A symbolic interactionist perspective. *Academy of Management Review, 38*(2), 181–205.

Eliot, T. (1943). *Four quartets.* Faber and Faber.

Gordon, R. (2016). *The rise and fall of American growth.* Princeton University Press.

Gruver, G., Young, M., & Fulghum, P. (2012). *A practical approach to large-scale agile development: How HP transformed LaserJet FutureSmart firmware.* Addison-Wesley Professional.

I Teach University. (2016, March 21). *www.iteachu.com.* Retrieved from www.iteachu.com, https://iteachu.uaf.edu

Kearns Goodwin, D. (2018). *Leadership in turbulent times.* Simon & Schuster.

Keates, C. (2018, Sept 10). *The Five Cs of Effective Communication.* Retrieved from Forbes: www.forbes.com

Kelly, S., & Allison, M. A. (1999). *The complexity advantage: How the science of complexity can help your business achieve peak performance*. McGraw-Hill.

Kirkmann, A. (2006, Jan). Contemporary linguistic theories of humour. *Folklore, 33*, 27–57. https://doi.org/10.7592/FEJF2006.33.kriku

KnowledgeHut Solutions Private Limited. (2011–23). *knowledgehut.com*. Retrieved from knowledgehut.com: www.knowledgehut.com

Kotter, J. (2012). *Leading change*. Harvard University Press.

Levy, D. (2000). *Applications and limitations of complexity theory in organization theory and strategy*. Computer Science. https://doi.org/10.4324/9781482270259-3

Lewin, A. Y. (1999). Application of complexity theory to organization science. *Organization Science, 10*(3), 215.

Lewin, R., & Regine, B. (2001). *Weaving complexity and business: Engaging the soul at work*. Cengage Learning.

Lewis, R. (1994). From chaos to complexity: Implications for organizations. *Executive Development, 7*(4), 16–17.

Lippmann, W. (1914). *Drift and mastery: An attempt to diagnose the current unrest*. Mitchell Kennerley.

ProductPlan. (2023). *www.productplan.com*. Retrieved from ProductPlan web site: www.productplan.com

Regine, B., & Lewin, R. (2000). Leading at the edge: How leaders influence complex systems. *Emergence, 2*, 23–52.

Rosenhead, J., Franco, L. A., Grint, K., & Friedland, B. (2019). Complexity theory and leadership practice: A review, a critique, and some recommendations. *The Leadership Quarterly, 30*, 1–25.

Scaled Agile. (2021, March). *scaledagileframework*. Retrieved from www.scaledagileframework.com

Schneider, A., Wickert, C., & Marti, E. (2017). Reducing complexity by creating complexity: A systems theory perspective on how organisations respond to their environments. *Journal of Management Studies, 54*, 182–208.

Schweiger, S., Muller, B., & Guttel, W. H. (2020). Barriers to leadership: Why is it so difficult to abandon the hero? *Leadership, 16*(4), 411–433. https://doi.org/10.1177/1742715020935742

Science History Publications/USA & The Nobel Museum. (2007). *Cultures of creativity: Birth of a 21st century museum*. Watson Publishing International LLP.

Simon, H. (1996). *The sciences of the artificial*. MIT Press.

Sinek, S. (2011). *Start with why: How great leaders inspire everyone to take action*. Penguin Books.

Smith, W., Erez, M. J., Lewis, M., & Tracey, P. (2017). Adding complexity to theories of paradox, tensions, and dualities of innovation and change: Introduction to the special issue on paradox, tensions, and dualities of innovation and change. *Organization Studies, 38*, 303–317.

Spector, B. (2014). Using history ahistorically: Presentism and the tranquility fallacy. *Management and Organizational History, 9*, 305–313.

Stacey, R. (1996). Emerging strategies for a chaotic environment. *Long Range Planning, 29*(2), 182–189.

The University of Texas Permian Basin. (2022, July 19). *How much of communication is nonverbal?*. Retrieved from The University of Texas Permian Basin: https://online.utpb.edu/about-us/article/communications/how-much-of-communication-is-nonverbal

Thompson, E. (1966). *The making of the English working class*. Penguin Books.

Tourish, D. (2019). Is complexity leadership theory complex enough? A critical appraisal, some modifications and suggestions for further research. *Organization Studies, 40*, 219–238.

Tourish, D. (2020). *National Leadership Centre*. Retrieved from National Leadership Centre: https://assets.publishing.service.gov.uk/government/uploads/system/uploads/attachment_data/file/926853/NLC-thinkpiece-Effective-Leadership-TOURISH.pdf

Tsoukas, H. (2017). Don't simplify, complexify: From disjunctive to conjunctive theorizing in organization and management studies. *Journal of Management Studies, 54*, 132–153.

Uhl-Bien, M., & Arena, M. (2017). Complexity leadership: Enabling people and organizations for adaptability. *Organizational Dynamics, 46*, 9–20.

Uhl-Bien, M., & Arena, M. (2018). Leadership for organizational adaptability: A theoretical synthesis and integrative framework. *Leadership Quarterly.* https://doi.org/10.1016/j.leaqua.2017.12.009

Zimmerman, B., Lindberg, C., & Plsek, P. (2008). *Edgeware: Lessons from complexity science for health care leaders* (2nd ed.). VHA, Incorporated.

Index

A
Ackenhusen, H., 70, 71, 102, 105, 131, 147, 149
Agile Manifesto, 14, 46–47
Alice in Wonderland, 139
Allison, M.A., 85, 173
Alvesson, M., 27
Amazon, 84, 88
Arena, M., 6, 17, 24, 70, 85, 171
Arora, U., 53, 108, 123
Automation, 148

B
Backlander, G., 77, 82, 172
Betz, K., 106, 162, 163, 171
Bion, 62
Bisociation, theory of, 23
Bloom, N., 44, 123, 124
Boal, K.B., 25, 94
Boundaries, 43, 64, 155
Brendzal, M., 51, 52, 54, 69, 71, 161
Burger, A., 16

C
Cares, 115, 139
Carroll, L., xiii, 139
Churchill, W., 109
Clarity, 64, 65, 76, 77, 84, 115, 122
    importance of, 61
Collinson, D., 7, 22, 28, 62
Collinson, M., 28
Combinatorial play, 23
Communications
    effective, 123, 127
    efficient, 122–125, 127
    face to face, 122–126

Conjunctive, xii, 26, 29, 52, 73, 104
    conjunctive thinking, 23
Conjunctive leadership, 23–24
    defined, 23
    what does the leader do, 24–25
    why it matters, 24
Conjunctive thinking, 23
Coursera, 122
Credibility, 142
Crowley, S., 22

D
Dawson, J.F., 172
Delivers, 64, 67, 142
Demo, 69–72, 78, 84, 106, 117,
    141, 166
    defined, 69
    purpose of, 69
Denis, J.-L., 25
Denning, S., 39, 50, 153
Dialogues, 28, 77–80, 82, 84, 85, 94, 102, 106,
    109, 111, 112, 117–119, 126, 137, 143,
    151, 160
    meaningful, 77
Dionysiou, D.D., 26
Diversity, 38
Drake, D., ix, 22, 38

E
Edmondson, A.C., 172
Einstein, A., 23, 114, 158
Electrification, 7
Eliot, T.S., 24, 114
Emerges, 83, 159, 160
Emerson, R., 155

Environments, 92–96, 99, 100, 102, 105, 114, 119, 122, 126, 148, 149, 164, 171
 physical, 92
Erez, M.J., 8

F
Face to face, 125
Failure, 54–56, 113–114
Farid, H., 54, 71, 101, 105, 107, 131, 133, 154–156, 164, 166
Feynman, R., 173
Franco, L.A., 9, 25, 38, 83, 94
French, R., 62
Friedland, B., 9, 25, 38, 83, 94
Fulghum, P., 27, 133, 140, 159, 160, 162

G
Generative, xii, 52, 53, 56, 63, 70, 73, 77, 79, 83, 89, 94, 112, 117, 118, 121, 125–127, 136, 137, 139, 151
 and relationships, 82
Generative leadership, 28
 defined, 28
 what does the leader do, 29–30
 why does it matter, 28–29
Gijselaers, W., 172
Gordon, R., 7
Grint, K., 9, 25, 38, 83, 94
Gruver, G., 27, 133, 140, 159, 160, 162
Guttel, W.H., 25, 28

H
Han, R., 123, 124
Harczak, H., 107
Hituvalli, D., 15, 77, 78, 132, 138–140
Hoven, M., 172
Howard, R., 68, 142
Humility, 24
Hybrid, 123, 124, 126
Hybrid work model, 41–44

I
Inspired, 90
Irizarry, M., xiv, xv, 22, 23, 25, 29, 41, 100
I Teach University, 23

K
Kearns Goodwin, D., 23
Keates, C., 122
Kelly, S., 85, 173

King, A., 42, 52, 62, 88, 147, 155
Kirkmann, A., 23
Koeslag-Kreunen, M., 172
Koestler, A., 23
Kotter, J.P., 17, 83
Krehbiel, M., 171

L
Langley, A., 25
Leader, 88
Leadership, xii, 23–28, 171
 definitions, 22
Leonard-Barton, 171
Levy, D., 91
Lewin, A.Y., 83, 105
Lewin, R., 82, 85
Lewis, M., 8
Lewis, R., 82
Liang, J., 44, 123, 124
Lindberg, C., 16, 17, 83
Linux Foundation, 149
Lippmann, W., 6–8, 31
Listen, xiii–xv, 38, 68–71, 123, 156
Listening, 97
Lowell, D., 24, 29, 41, 43, 44, 54, 61, 62, 92, 93, 99, 100, 124

M
Mander, J., 42, 43, 60, 81, 82, 104, 142, 165, 166
March, 171
Mentors, 116, 118
 ending the mentoring relationship, 118
 mentoring tenured employees, 118
 reverse mentoring, 118
 traditional mentoring, 116–118
Michaels, A., ix
Michelangelo, xvi, 11
Mintzberg, 8
Motivation
 individuals, 87
 motivated individuals, 87, 89
Muller, B., 25, 28

O
O'Leary, C., ix, 24
Obeidat, Muhannad, 16, 61, 69, 70, 171

P
Peters, T., 8
Plsek, P., 16, 17, 83

Index

Presentism, 8
Principles, xii–xvi, 14, 50, 59, 67, 75, 87, 121–126, 133, 136, 145, 153, 159, 170
Processual, xii, 52, 73, 80, 81, 115, 151
Processual leadership, 25
  defined, 25
  what does the leader do, 26–28
  why it matters, 25–26
ProductPlan, 17
Purposes, 8, 37, 122

R
Reflect, 170
Reflection, 171, 173
Regine, B., 82, 85
Relationships, 94
Risk, 54–56
Robeson, S., 56
Roosevelt, F.D., 23
Rosenhead, J., 9, 25, 38, 83, 94

S
Schippers, M.C., 172
Schneider, A., 83
Schultz, P.L., 25, 94
Schweiger, S., 25, 28
Sergi, V., 25
Simon, H., 17
Simplicity, 153
Simpson, P., 62
Sinek, S., 61
Smith, W., 8
Social construct, 22–23
Software, 150
Software development, 13, 46
Spector, B., 8
Stacey, R., 85
Stewart, C., 22, 38
Supports
  at work, 96
Sveningsson, S., 27

T
Talent, 99–115
  attracting, 99–100
  developing, 104–110
  development plans, 108
  empowering, 104
  equipping with information, 104
  equipping with resources, 104
  equipping with tools, 103–104
  getting feedback, 111
  giving feedback, 111–113
  hiring, 100
  identifying, 100–103
  incentives, 114–115
  making mistakes, 105
  mentoring, 115–118
Taylorism, 31
Therivel, Laurent "LT", 27, 37, 45, 115
Thompson, E., 9, 36
Thoreau, H., 155
Tiselius, A., 28
Tourish, D., 6–8, 23, 26–29, 82
Tracey, P., 8
Trust, 88, 98–99, 103
Tsoukas, H., 23, 25, 26, 104, 160, 172

U
Uhl-Bien, M., 6, 17, 24, 70, 85, 171

V
Valuable, 50, 131
Values, 27, 51, 52, 82, 96, 154, 171
Van den Bossche, P., 172
Van der Klink, M., 172
VUCA, 8

W
Wallin, C., ix, 22, 96, 161, 162
Walter, T., 37
West, M.A., 172
Whitman, W., 35
Wiesel, E., 16
Wooden, J., 97
Working software, 129–131

Y
Young, J., 170
Young, M., 27, 133, 140, 159, 160, 162

Z
Zimmerman, B., 16, 17

GPSR Compliance

The European Union's (EU) General Product Safety Regulation (GPSR) is a set of rules that requires consumer products to be safe and our obligations to ensure this.

If you have any concerns about our products, you can contact us on

ProductSafety@springernature.com

In case Publisher is established outside the EU, the EU authorized representative is:

Springer Nature Customer Service Center GmbH
Europaplatz 3
69115 Heidelberg, Germany